Tucholsky Wagner Zola Scott Sydow Schlegel
Turgenev Wallace Fonatne Freud
Twain Walther von der Vogelweide Fouqué Friedrich II. von Preußen
Weber Freiligrath Frey
Fechner Fichte Weiße Rose von Fallersleben Kant Ernst Richthofen Frommel
Engels Fielding Hölderlin
Fehrs Faber Flaubert Eichendorff Tacitus Dumas
Maximilian I. von Habsburg Fock Eliasberg Ebner Eschenbach
Feuerbach Ewald Eliot Zweig Vergil
Goethe Elisabeth von Österreich London
Mendelssohn Balzac Shakespeare Dostojewski Ganghofer
Trakl Stevenson Lichtenberg Rathenau Doyle Gjellerup
Mommsen Thoma Tolstoi Lenz Hambruch Droste-Hülshoff
Dach Verne von Arnim Hägele Hanrieder Hauff Humboldt
Reuter
Karrillon Garschin Rousseau Hagen Hauptmann Gautier
Damaschke Defoe Hebbel Baudelaire
Descartes
Wolfram von Eschenbach Darwin Dickens Schopenhauer Hegel Kussmaul Herder
Bronner Melville Grimm Jerome Rilke George
Campe Horváth Aristoteles Bebel Proust
Bismarck Vigny Barlach Voltaire Federer Herodot
Gengenbach Heine
Storm Casanova Tersteegen Gilm Grillparzer Georgy
Chamberlain Lessing Langbein Gryphius
Brentano Lafontaine
Strachwitz Claudius Schiller Kralik Iffland Sokrates
Katharina II. von Rußland Bellamy Schilling
Gerstäcker Raabe Gibbon Tschechow
Löns Hesse Hoffmann Gogol Wilde Gleim Vulpius
Luther Heym Hofmannsthal Klee Hölty Morgenstern Goedicke
Roth
Luxemburg Heyse Klopstock Puschkin Homer Kleist Mörike
Machiavelli La Roche Horaz Musil
Navarra Aurel Musset Kierkegaard Kraft Kraus Moltke
Nestroy Marie de France Lamprecht Kind Kirchhoff Hugo
Laotse Ipsen Liebknecht
Nietzsche Nansen Ringelnatz
Marx Lassalle Gorki Klett Leibniz
von Ossietzky May vom Stein Lawrence Irving
Petalozzi Knigge
Platon Pückler Michelangelo Kock Kafka
Sachs Poe Liebermann Korolenko
de Sade Praetorius Mistral Zetkin

The publishing house tredition has created the series **TREDITION CLASSICS**. It contains classical literature works from over two thousand years. Most of these titles have been out of print and off the bookstore shelves for decades.

The book series is intended to preserve the cultural legacy and to promote the timeless works of classical literature. As a reader of a **TREDITION CLASSICS** book, the reader supports the mission to save many of the amazing works of world literature from oblivion.

The symbol of **TREDITION CLASSICS** is Johannes Gutenberg (1400 – 1468), the inventor of movable type printing.

With the series, tredition intends to make thousands of international literature classics available in printed format again – worldwide.

All books are available at book retailers worldwide in paperback and in hardcover. For more information please visit: www.tredition.com

tredition was established in 2006 by Sandra Latusseck and Soenke Schulz. Based in Hamburg, Germany, tredition offers publishing solutions to authors and publishing houses, combined with worldwide distribution of printed and digital book content. tredition is uniquely positioned to enable authors and publishing houses to create books on their own terms and without conventional manufacturing risks.

For more information please visit: www.tredition.com

The Compleat Cook Expertly Prescribing the Most Ready Wayes, Whether Italian, Spanish or French, for Dressing of Flesh and Fish, Ordering Of Sauces or Making of Pastry

W. M.

Imprint

This book is part of the TREDITION CLASSICS series.

Author: W. M.
Cover design: toepferschumann, Berlin (Germany)

Publisher: tredition GmbH, Hamburg (Germany)
ISBN: 978-3-8491-6651-9

www.tredition.com
www.tredition.de

Copyright:
The content of this book is sourced from the public domain.

The intention of the TREDITION CLASSICS series is to make world literature in the public domain available in printed format. Literary enthusiasts and organizations worldwide have scanned and digitally edited the original texts. tredition has subsequently formatted and redesigned the content into a modern reading layout. Therefore, we cannot guarantee the exact reproduction of the original format of a particular historic edition. Please also note that no modifications have been made to the spelling, therefore it may differ from the orthography used today.

THE COMPLEAT COOK.

Expertly prescribing the most ready wayes,

Whether, { *Italian*, { *Spanish*, { or *French*

For dressing of *Flesh*, and *Fish*, Ordering of *Sauces* or making
OF
PASTRY.

THE COMPLEAT COOK:

Expertly prescribing the most ready wayes, whether *Italian, Spanish,* or *French,* for dressing of *Flesh* and *Fish,* &c.

To make a Posset, the Earle of Arundels Way..

Take a quart of Creame, and a quarter of a Nutmeg in it, then put it on the fire, and let it boyl a little while, and as it is boyling take a Pot or Bason, that you meane to make your Posset in, and put in three spoonfuls of Sack, and some eight of Ale, and sweeten it with Sugar, then set it over the coles to warm a little while, then take it off and let it stand till it be almost cool, then put it into the Pot or Bason and stir it a little, and let it stand to simper over the fire an hour or more, for the longer the better.

To boyle a Capon larded with Lemons.

Take a fair Capon and truss him, boyl him by himselfe in faire water with a little small Oat-meal, then take Mutton Broth, and half a pint of White-wine, a bundle of Herbs, whole Mace, season it with Verjuyce, put Marrow, Dates, season it with Sugar, then take preserved Lemons and cut them like Lard, and with a larding pin, lard in it, then put the capon in a deep dish, thicken your broth with Almonds, and poure it on the Capon.

To Bake Red Deere.

Parboyl it, and then sauce it in Vinegar then Lard it very thick, and season it with Pepper, Ginger and Nutmegs, put it into a deep Pye with good store of sweet butter, and let it bake, when it is baked, take a pint of Hippocras, halfe a pound of sweet butter, two or three Nutmeg, little Vinegar, poure it into the Pye in the Oven and let it lye and soake an hour, then take it out, and when it is cold stop the vent hole.

To make fine Pan-cakes fryed without Butter or Lard.

Take a pint of Cream, and six new laid Egs, beat them very well together, put in a quarter of a pound of Sugar, and one Nutmeg or a little beaten Mace (which you please) and so much flower as will thicken almost as much as ordinarily Pan-cake batter; your Pan must be heated reasonably hot & wiped with a clean Cloth, this done put in your Batter as thick or thin as you please.

To dresse a Pig the French manner.

Take it and spit it, & lay it down to the fire, and when your Pig is through warme, skin her, and cut her off the Spit as another Pig is, and so divide it in twenty peeces more or lesse as you please; when you have so done, take some White-wine and strong broth, and stew it therein, with an Onion or two mixed very small, a little Time also minced with Nutmeg sliced and grated Pepper, some Anchoves and Elder Vinegar, and a very little sweet Butter, and Gravy if you have it, so Dish it up with the same Liquor it is stewed in, with French Bread sliced under it, with Oranges and Lemons.

To make a Steake pye, with a French Pudding in the Pye.

Season your Steaks with Pepper & Nutmegs, and let it stand an hour in a Tray then take a piece of the leanest of a Legg of Mutton and mince it small with Suet and a few sweet herbs, tops of young Time, a branch of Penny-royal, two or three of red Sage, grated bread, yolks of Eggs, sweet Cream, Raisins of the Sun; work altogether like a Pudding, with your hand stiff, and roul them round like Bals, and put them into the Steaks in a deep Coffin, with a piece of sweet Butter; sprinkle a little Verjuyce on it, bake it, then cut it up and roul Sage leaves and fry them, and stick them upright in the wals, and serve your Pye without a Cover, with the juyce of an Orange or Lemon.

An excellent way of dressing Fish.

Take a piece of fresh Salmon, and wash it clean in a little Vinegar and water, and let it lie a while in it, then put it into a great Pipkin with a cover, and put to it some six spoonfuls of water and four of Vinegar, and as much of white-wine, a good deal of Salt a handful of sweet herbs, a little white Sorrel, a few Cloves, a little stick of Cinamon, a little Mace; put all these in a Pipkin close, and set it in a Kettle of seething water, and there let it stew three hours.

You may do Carps, Eeles, Trouts, &c. this way, and they Tast also to your mind.

To fricate Sheeps-feet.

Take Sheeps-feet, slit the bone, and pick them very clean, then put them in a Frying-pan, with a Ladlefull of strong Broth, a piece of Butter, and a little Salt, after they have fryed a while, put to them a little Parsley, green Chibals, a little young Speremint and Tyme, all shred very small, and a little beaten Pepper; when you think they are fryed almost enough, have a lear made for them with the yolks of two or three Eggs, some Gravy of Mutton, a little Nutmegg, and juyce of a Lemon wrung therein, and put this lear to the Sheeps feet as they fry in the Pan, then toss them once or twice, and put them forth into the Dish you mean to serve them in.

To fricate Calves Chaldrons.

Take a Calves Chaldron, after it is little more then half boyled, and when it is cold, cut it into little bits as big as Walnuts; season it with beaten Cloves, Salt, Nutmeg, Mace, and a little Pepper, an Onion, Parsley, and a little Tarragon, all shred very small, then put it into a frying-pan, with a Ladle-full of strong broth, and a little piece of sweet Butter, so fry it; when it is fryed enough, have a little lear made with the Gravy of Mutton, the juyce of a Lemon and Orange, the yolks of three or four Eggs, and a little Nutmeg grated therein; put all this to your Chaldrons in the Pan, Toss your Fricat two or three times, then dish it, and so serve it up.

To Fricate Champigneons.

Make ready your champigneons as you do for stewing, and when you have poured away the black liquor that comes from them, put your champigneons into a Frying pan with a piece of sweet Butter, a little Parsley, Tyme, sweet Marjoram, a piece of Onion shred very small, a little Salt and fine beaten Pepper, so fry them till they be enough, so have ready the lear abovesaid, and put it to the champigneons whilst they are in the Pan, toss them two or three times, put them forth and serve them.

To make buttered Loaves.

Take the yolks of twelve Eggs, and six whites, and a quarter of a pint of yeast, when you have beaten the Eggs well, strain them with the yeast into a Dish, then put to it a little Salt, and two rases of Ginger beaten very small, then put flower to it till it come to a high Past that will not cleave, then you must roule it upon your hands and afterwards put it into a warm Cloath and let it lye there a quarter of an hour, then make it up in little Loaves, bake; against it is baked prepare a pound and a half of Butter, a quarter of a pint of white wine, and halfe a pound of Sugar; This being melted and beaten together with it, set them into the Oven a quarter of an hour.

To murine Carps, Mullet, Gurnet, Rochet, or Wale, &c.

Take a quart of water to a Gallon of Vinegar, a good handful of Bay-leaves, as much Rosemary, a quarter of a pound of Pepper beaten; put all these together, and let it seeth softly, and season it with a little Salt, then fry your Fish with frying Oyle till it be enough, then put in an earthen Vessell, and lay the Bay-leaves and Rosemary between and about the Fish, and pour the Broth upon it, and when it is cold, cover it, &c.

To make a Calves Chaldron Pye.

Take a Calves Chaldron, half boyl it, and cool it; when it is cold mince it as small as grated bread, with halfe a pound of Marrow; season it with Salt, beaten Cloves, Mace, Nutmeg a little Onion, and some of the outmost rind of a Lemon minced very small, and wring in the juyce of halfe a Lemon, and then mix all together, then make a piece of puff Past, and lay a leaf therof in a silver Dish of the bigness to contain the meat, then put in your meat, and cover it with another leaf of the same Past, and bake it; and when it is baked take it out, and open it, and put in the juyce of two or three Oranges, stir it well together, then cover it againe and serve it. Be sure none of your Orange kernels be among your Pye-meat.

To make a Pudding of a Calves Chaldron.

Take your Chaldron after it is half boyled and cold, mince it as small as you can with half a pound of Beef Suet, or as much Marrow, season it with a little Onion, Parsley, Tyme, and the outmost rind of a piece of Lemon, all shred very small, Salt, beaten Nutmeg, Cloves and mace mixed together, with the yolks of four or five Eggs, and a little sweet Cream; then have ready the great Gutts of a Mutton scraped and washed very clean; let your Gutt have lain in white-wine and Salt halfe a day before you use it; when your meat is mixed and made up somewhat stiff put it into the Sheeps-gutt, and so boyl it, when it is boyled enough, serve it to the Table in the Gutt.

To make a Banbury Cake.

Take a peck of pure Wheat-flower, six pound of Currans, half a pound of Sugar, two pound of Butter, halfe an ounce of Cloves and Mace, a pint and a halfe of Ale-yeast, and a little Rose-water; then boyle as much new-milk as will serve to knead it, and when it is almost cold, put into it as much Sack as will thicken it, and so work it all together before a fire, pulling it two or three times in pieces, after make it up.

To make a Devonshire White-pot.

Take a pint of Cream and straine four Eggs into it, and put a little Salt and a little sliced Nutmeg, and season it with Sugar somewhat sweet; then take almost a penny Loaf of fine bread sliced very thin, and put it into a Dish that will hold it, the Cream and the Eggs being put to it; then take a handfull of Raisins of the Sun being boyled, and a little sweet Butter, so bake it.

To make Rice Cream.

Take a quart of Cream, two good handfuls of Rice-flower, a quarter of a pound of Sugar and flower beaten very small, mingle your Sugar and flower together, put it into your Cream, take the yolk of an Egg, beat it with a spoonfull or two of Rose-water, then put it to the Cream, and stir all these together, and set it over a quick fire, keeping it continually stirring till it be as thick as water-pap.

To make a very Good Great Oxford-shire Cake.

Take a peck of flower by weight, and dry it a little, & a pound and a halfe of Sugar, one ounce of Cinamon, half an ounce of Nutmegs, a quarter of an ounce of Mace and Cloves, a good spoonfull of Salt, beat your Salt and Spice very fine, and searce it, and mix it with your flower and Sugar; then take three pound of butter and work it in the flower, it will take three hours working; then take a quart of Ale-yeast, two quarts of Cream, half a pint of Sack, six grains of Amber-greece dissolved in it, halfe a pint of Rosewater, sixteen Eggs, eight of the Whites, mix these with the flower, and knead them well together, then let it lie warm by your fire till your Oven be hot, which must be little hotter then for manchet; when you make it ready for your Oven, put to your Cake six pound of Currans, two pound of Raisins, of the Sun stoned and minced, so make up your Cake, and set it in your oven stopped close; it wil take three houres a baking; when baked, take it out and frost it over with the white of an Egge and Rosewater, well beat together, and strew fine Sugar upon it, and then set it again into the Oven, that it may Ice.

To make a Pumpion Pye.

Take about halfe a pound of Pumpion and slice it, a handfull of Tyme, a little Rosemary, Parsley and sweet Marjoram slipped off the stalks, and chop them smal, then take Cinamon, Nutmeg, Pepper, and six Cloves, and beat them; take ten Eggs and beat them; then mix them, and beat them altogether, and put in as much Sugar as you think fit, then fry them like a froiz; after it is fryed, let it stand till it be cold, then fill your Pye, take sliced Apples thinne round wayes, and lay a row of the Froiz, and a layer of Apples with Currans betwixt the layer while your Pye is fitted, and put in a good deal of sweet butter before you close it; when the Pye is baked, take six yolks of Eggs, some white-wine or Verjuyce, & make a Caudle of this, but not too thick; cut up the Lid and put it in, stir them well together whilst the Eggs and Pumpions be not perceived, and so serve it up.

To make the best Sausages that ever was eat.

Take a leg of young Pork, and cut of all the lean, and shred it very small, but leave none of the strings or skins amongst it, then take two pound of Beef Suet, and shred it small, then take two handfuls of red Sage, a little Pepper and Salt, and Nutmeg, and a small piece of an Onion, chop them altogether with the flesh and Suet; if it is small enough, put the yolk of two or three Eggs and mix altogether, and make it up in a Past if you will use it, roul out as many pieces as you please in the form of an ordinary Sausage, and so fry them, this Past will keep a fortnight upon occasion.

To boyle a Fresh Fish.

Take a Carp, or other, & put them into a deep Dish, with a pint of white-wine, a large Mace, a little Tyme, Rosemary, a piece of sweet Butter, and let him boyle between two dishes in his owne blood, season it with Pepper and Verjuyce, and so serve it up on Sippets.

To make Fritters.

Take halfe a pint of Sack, a pint of Ale, some Ale-yeast, nine Eggs, yolks and whites, beat them very well, the Egg first, then altogether, put in some Ginger, and Salt, and fine flower, then let it stand an houre or two; then shred in the Apples; when you are ready to fry them, your suet must be all Beef-suet, or halfe Beef, and halfe Hoggs-suet tryed out of the leafe.

To make Loaves of Cheese-Curds.

Take a Porringer full of Curds, and four Eggs, whites, and yolks, and so much flower as will make it stiff, then take a little Ginger, Nutmeg, & some Salt, make them into loaves and set them into an oven with a quick heat; when they begin to change Colour take them out, and put melted Butter to them, and some Sack, and good store of Sugar, and so serve it.

To make fine Pies after the French fashion.

Take a pound and half of Veale, two pound of suet, two pound of great Raisins stoned, half a pound of Prunes, as much of Currans, six Dates, two Nutmegs, a spoonfull of Pepper, an ounce of Sugar, an ounce of Carrawayes, a Saucer of Verjuyce, and as much Rose-water, this will make three fair Pyes, with two quarts of flower, three yolks of Egges, and halfe a pound of Butter.

A Singular Receit for making a Cake.

Take halfe a peck of flower, two pound of Butter, mingle it with the flower, three Nutmegs, & a little Mace, Cinamon, Ginger, halfe a pound of Sugar, leave some out to strew on the top, mingle these well with the flower and Butter, five pound of Currans well washed, and pickt, and dryed in a warm Cloth, a wine pint of Ale yeast, six Eggs, leave out the whites, a quart of Cream boyled and almost cold againe: work it well together and let it be very lith, lay it in a warm Cloth, and let it lye half an hour against the fire. Then make it up with the white of an Egg, a little Butter, Rosewater and

Sugar; Ice it over and put it into the Oven, and let it stand one whole hour and a half.

To make a great Curd Loaf.

Take the Curds of three quarts of new milk clean whayed, and rub into them a little of the finest flower you can get, then take half a race of Ginger, and slice it very thin, and put it into your Curds with a little Salt, then take halfe a pint of good Ale Yeast and put to it, then take ten Eggs, but three of the Whites, let there be so much flower as will make it into a reasonable stiff Past, then put it into an indifferant hot cloth, and lay it before the fire to rise while your Oven is heating, then make it up into a Loaf, and when it is baked, cut up the top of the Loaf, and put in a pound and a half of melted Butter, and a good deale of Sugar in it.

To make buttered Loaves of Cheese-curds.

Take three quarts of new Milk, and put in as much Rennet as will turn, take your Whay clean away, then breake your Curds very small with your hands, and put in six yolks of Eggs, but one white; an handfull of grated bread, an handfull of Flower, a little Salt mingled altogether; work it with your hand, roul it into little Loaves, then set them in a Pan buttered, then beat the yolk of an Egg with a little Beer, and wipe them over with a feather, then set them in the Oven as for Manchet, and stop that close three quarters of an hour, then take halfe a pound of butter three spoonfuls of water, a Nutmeg sliced thin, a little Sugar, set it on the fire, stir it till it be thick; when your Loaves are baked, cut off the tops and butter them with this Butter, some under, some over, and strow some Sugar on them.

To make Cheese-loaves.

Grate a Wheat-Loafe, and take as much Curd as bread, to that put eight yolks of Eggs and four whites, and beat them very well, then take a little Cream but let it be very thick, put altogether, and make them up with two handfuls of flower, the Curds must be made of new milk and whayed very dry, you must make them like little Loaves and bake them in an Oven; and being baked cut them up, and have in readinesse some sweet Butter, Sugar, Nutmeg sliced and mingled together, put it into the Loaves, and with it stir the Cream well together, then cover them again with the tops, and serve them with a little Sugar scraped on.

To make Puff.

Take four pints of new milke, rennet, take out all the Whay very clean, and wring it in a dry Cloth, then strain it in a wooden Dish till they become as Cream, then take the yolks of two Egges, and beat them and put them to the Curds, and leave them with the Curds, then put a spoonfull of Cream to them, and if you please halfe a spoonfull of Rose-water, and as much flower beat in it as will make it of an indifferent stiffnesse, just to roul on a Plate, then take off the Kidney of Mutton suet and purifie it, and fry them in it, and serve them with Butter, Rose-water and Sugar.

To make Elder Vinegar.

Gather the flowers of Elder, pick them very clean, and dry them in the Sun on a gentle heat, and take to every quart of Vinegar a good handfull of flowers and let it stand to Sun a fortnight, then strain the Vinegar from the flowers, and put it into the barrell againe, and when you draw a quart of Vinegar, draw a quart of water, and put it into the Barrell luke warme.

To make good Vinegar.

Take one strike of Malt, and one of Rye ground, and mash them together, and take (if they be good) three pound of Hops, if not four pound; make two Hogs-heads of the best of that Malt and Rye, then lay the Hogs-head where the Sunne may have power over them, and when it is ready to Tun, fill your hogs-heads where they lye, then let them purge cleer and cover them with two flate stones, and within a week after when you bake, take two wheat loaves hot out of the Oven, and put into each hogs-head a loaf, you must use this foure times, you must brew this in *Aprill,* and let it stand till *June,* then draw them clearer, then wash the Hogs-heads cleane, and put the beer in again; if you will have it Rose-vinegar, you must put in a strike and a half of Roses; if Elder-vinegar, a peck of the flowers; if you will have it white, put no thing in it after it is drawn, and so let it stand till *Michaelmas*; if you will have it coloured red, take four gallons of strong Ale as you can get, and Elder berries picked a few full clear, and put them in your pan with the Ale, set them ouer the fire till you guesse that a pottle is wasted, then take if off the fire, and let it stand till it be store cold, and the next day strain it into the Hogs-head, then lay them in a Cellar or buttery which you please.

To make a Coller of Beef.

Take the thinnest end of a coast of beef, boyl it and lay it in Pump-water, and a little salt, three dayes shifting it once every day, and the last day put a pint of Claret Wine to it, and when you take it out of the water, let it lye two or three hours a drayning, then cut it almost to the end in three slices, then bruise a little Cochinell and a very little Allum, and mingle it with the Claret-wine, and colour the meat all over with it, then take a dozen of Anchoves, wash them and bone them, and lay them into the Beef, and season it with Cloves, Mace, and Pepper, and two handfuls of salt, and a little sweet Marjoram and Tyme, and when you make it up, roul the innermost slice first, and the other two upon it, being very wel seasoned every where, and bind it hard with Tape, then put it into a stone-pot, something bigger then the Coller, and pour upon it a pint of Claret-wine, and halfe a pint of wine-vinegar, a sprig of Rosemary, and a few Bay-leave and bake it very well; before it is quite cold, take it out of the Pot, and you may keep it dry as long as you please.

To make an Almond Pudding.

Take two or three French-Rowles, or white penny bread, cut them in slices, and put to the bread as much Cream as wil cover it, put it on the fire till your Cream and bread be very warm, then take a ladle or spoon and beat it very well together, put to this twelve Eggs, but not above foure whites, put in Beef Suet, or Marrow, according to your discretion, put a pretty quantity of Currans and Raisins, season the Pudding with Nutmeg, Mace, Salt, and Sugar, but very little flower for it will make it sad and heavy; make a piece of puff past as much as will cover your dish, so cut it very handsomely what fashion you please; Butter the bottom of your Dish, put the pudding into the Dish, set it in a quick Oven, not too hot as to burne it, let it bake till you think it be enough, scrape on Sugar and serve it up.

To boyle Cream with French Barly.

Take the third part of a pound of French Barley, wash it well with fair water, and let it lie all night in fair water, in the morning set two skillets on the fire with faire water, and in one of them put your Barley, and let it boyle till the water look red, then put the water from it, and put the Barley into the other warme water, thus boyl it and change with fresh warm water till it boyl white, then strain the water clean from it, then take a quart of Creame, put into it a Nutmeg or two quartered, a little large Mace and some Sugar, and let it boyl together a quarter of an hour, and when it hath thus boyled put into it the yolks of three or foure Eggs, well beaten with a little Rose-water, then dish it forth, and eat it cold.

To make Cheese-Cakes.

Take three Eggs and beat them very well, and as you beat them, put to them as much fine flower as will make them thick, then put to them three or four Eggs more, and beat them altogether; then take one quart of Creame, and put into it a quarter of a pound of sweet butter, and set them over the fire, and when it begins to boyle, put to it your Eggs and flower, stir it very well, and let it boyle till it be thick, then season it with Salt, Cinamon, Sugar, and Currans, and bake it.

To make a Quaking Pudding.

Take a pint and somewhat more of thick Creame, ten Egges, put the whites of three, beat them very well with two spoonfuls of Rosewater; mingle with your Creame three spoonfuls of fine flower, mingle it so well, that there be no lumps in it, put it altogether, and season it according to your Tast; Butter a Cloth very well, and let it be thick that it may not run out, and let it boyle for half an hour as fast as you can, then take it up and make Sauce with Butter, Rosewater and Sugar, and serve it up.

You may stick some blanched Almonds upon it if you please.

To Pickle Cucumbers.

Put them in an Earthen Vessel, lay first a Lay of Salt and Dill, then a Lay of Cucumbers, and so till they be all Layed, put in some Mace and whole pepper, and some Fennel-seed according to direction, then fill it up with Beer-Vinegar, and a clean board and a stone upon it to keepe them within the pickle, and so keep them close covered, and if the Vinegar is black, change them into fresh.

To Pickle Broom Buds.

Take your Buds before they be yellow on the top, make a brine of Vinegar and Salt, which you must do onely by shaking it together till the Salt be melted, then put in your Buds, and keepe stirred once in a day till they be sunk within the Vinegar, be sure to keep close covered.

To keep Quinces raw all the year.

Take some of the worst Quinces and cut them into small pieces, and Coares and Parings, boyle them in water, and put to a Gallon of water, some three spoonfuls of Salt, as much Honey; boyle these together till they are very strong, and when it is cold, put it into half a pint of Vinegar in a wooden Vessell or Earthen Pot; and take then as many of your best Quinces as will go into your Liquor, then stop them up very close that no Aire get into them, and they will keep all the yeare.

To make a Gooseberry Foole.

Take your Gooseberries, and put them in a Silver or Earthen Pot, and set it in a Skillet of boyling Water, and when they are coddled enough strain them, then make them hot again, when they are scalding hot, beat them very well with a good piece of fresh butter, Rose-water and Sugar, and put in the yolke of two or three Eggs; you may put Rose-water into them, and so stir it altogether, and serve it to the Table when it is cold.

To make an Otemeale Pudding.

Take a Porringer full of Oatmeale beaten to flower, a pint of Creame, one Nutmeg, four Eggs beaten, three whites, a quarter of a pound of Sugar, a pound of Beefe-suet well minced, mingle all these together and so bake it. An houre will bake it.

To make a green Pudding.

Take a penny loafe of stale Bread, grate it, put to halfe a pound of Sugar, grated Nutmeg, as much Salt as will season it, three quarters of a pound of beef-suet shred very small, then take sweet Herbs, the most of them Marigolds, eight Spinages: shred the Herbs very small, mix all well together, then take two Eggs and work them up together with your hand, and make them into round balls, and when the water boyles put them in, serve them with Rose-water, Sugar, and Butter or Sauce.

To make good Sausages.

Take the lean of a Legge of Pork, and four pound of Beefe-suet, or rather butter, shred them together very small, then season it with three quarters of an ounce of Pepper, and halfe an ounce of Cloves and Mace mixed together, as the Pepper is, a handfull of Sage when it is chopt small, and as much salt as you thinke will make them tast well of it; mingle all these with the meat, then break in ten Eggs, all but two or three of the whites, then temper it all well with your hands, and fill it into Hoggs gutts, which you must have ready for them; you must tye the ends of them like Puddings, and when you eat them you must boyle them on a soft fire; a hot will crack the skins, and the goodnesse boyle out of them.

To make Toasts.

Cut two penny Loaves in round slices and dip them in half a pint of Cream or cold water, then lay them abroad in a Dish, and beat three Eggs and grated Nutmegs, and Sugar, beat them with the Cream, then take your frying Pan and melt some butter in it, and wet one side of your Toasts and lay them in on the wet side, then pour in the rest upon them, and so fry them; send them in with Rosewater, butter and sugar.

Spanish Cream.

Put hot water in a bucket and go with it to the Milking, then poure out the Water, and instantly milke into it, and presently strain it into milk-Pans of an ordinary fulnesse, but not after an ordinary way for you must set your Pan on the ground and stand on a stool, and pour it forth that it may rise in bubbles with the fall; this on the morrow will be a very tough Cream, which you must take off with your Skimmer, and lay it in the Dish, laying upon laying; and if you please strew some sugar between them.

To make Clouted Cream.

Take foure quarts of Milke, one of Cream, six spoonfuls of Rose-water, put these together in a great Earthen Milke-Pan, & set it upon a fire of Charcoale well kindled, you must be sure the fire be not too hot; then let it stand a day and a night; and when you go to take it off, loose the edge of your Cream around about with a Knife, then take your board, and lay the edges that is left beside the board, cut into many pieces, and put them into the Dish first, and scrape some fine Sugar upon them, then take your board and take off your Cream as clean from the Milk as you can, and lay it upon your Dish, and if your Dish be little, there will be some left, the which you may put into what fashion you please, and scrape good store of Sugar upon it.

A good Cream

When you Churn Butter, take out six spoonfuls of Cream, just as it is to turne to Butter, that is, when it is a little frothy; then boyle good Cream as must as will make a Dish, and season it with Sugar, and a little Rose-water; when it is quite cold enough, mingle it very well with that you take out of the Churn, and so Dish it.

To make Piramidis Cream.

Take a quart of water, and six ounces of harts horn, and put it into a Bottle with Gum-dragon, and Gum-arabick, of each as much as a small Nut, put all this into the Bottle, which must be so big as will hold a pint more; for if it be full it will break; stop it very Close with a Cork, and tye a Cloth about it, put the Bottle into a pot of beef when it is boyling, and let it boyle three hours, then take as much Cream as there is Jelly, and halfe a pound of Almonds well beaten with Rose-water, so that you cannot discern what they be, mingle the Cream and the Almonds together, then strain it, and do so two or three times to get all you can out of the Almonds, then put jelly when it is cold into a silver Bason, and the Cream to it; sweeten it as you like, put in two or three grains of Musk and Amber-greece, set it over the fire, stirring it continually and skimming it, till it be seething hot, but let it not boyle, then put it into an old fashion drinking-Glasse, and let it stand till it is cold, and when you will use it, hold your Glass in a warm hand, and loosen it with a Knife, and whelm it into a Dish, and have in readinesse Pine Apple blown, and stick it all over, and serve it in with Cream or without as you please.

To make a Sack Cream.

Set a quart of Cream on the fire, when it is boyled, drop in a spoonfull of sack, and stir it well the while, that it curd not, so do till you have dropped in six spoonfuls, then season it with sugar, Nutmeg, and strong water.

To boyle Pigeons.

Stuffe the Pigeons with Parsley, and butter, and put them into an Earthen Pot, and put some sweet butter to them and let them boyle; take Parsley, Tyme, and Rosemary, chop them and put them to them; take some sweet butter, and put in withall some spinage, take a little gross Pepper and Salt, and season it withall, then take the yolk of an Egge and strain it with Verjuyce, and put to them, lay sippets in the Dish and serve it.

To make an Apple-Tansey.

Pare your Apples and cut them in thin round slices, then fry them in good sweet Butter, then take ten Eggs, sweet Cream, Nutmeg, Cinamon, Ginger, Sugar, with a little Rose-water, beat all these together, and poure it upon your Apples and fry it.

The French-Barly-Cream.

Take a quart of Cream, and boyle in a Porrenger of French-Barley, that hath been boyled in a nine waters, put in some large Mace and a little Cinamon, boyling it a quarter of an hour; then take two quarts of Almonds blanched, and beat it very small with Rose-water, or Orange-water, and some Sugar; and the Almonds being strained into the Liquor, put it over the fire, stirring it till it be ready to boyle; then take it off the fire, stirring it till it be halfe cold; then put to it two spoonfuls of Sack or White-Wine, and when it is cold, serve it in, remembring to put in some Salt.

To make a Chicken or Pigeon-Pye.

Take your Pigeons (if they be not very young) cut them into four quarters, one sweet-bread sliced the long way, that it may be thin, and the pieces not too big, one Sheeps tongue, little more then parboyl'd, and the skin puld off, and the tongue cut in slices, two or three slices of Veale, as much of Mutton, young chicken (if not little) quarter them, Chick-heads, Lark, or any such like, Pullets, Coxcombs, Oysters, Calves-Udder cut in pieces, good store of Marrow for seasoning, take as much Pepper and Salt as you think fit to season it slightly; good store of sweet Marjoram, a little Time and Lemon-Pill fine sliced; season it well with these Spices as the time of the year will afford; put in either of Chesnuts (if you put in Chesnuts they must first be either boyl'd or roasted) Gooseberries or Guage, large Mace will do well in this Pye, then take a little piece of Veale parboyl'd and slice it very fine, as much Marrow as meat stirred amongst it, then take grated Bread, as much as a quarter of the meat, four yolks of Eggs or more according to the stuffe you make, shred Dates as small as may be, season it with salt, but not too salt. Nutmeg as much as will season it, sweet Marjoram pretty store very small shred, work it up with as much sweet Creame as will make it up in little Puddings, some long, some round, so put as many of them in the Pye as you please; put therein two or three spoonfulls of Gravy of Mutton, or so much strong Mutton broth before you put it in the Oven, the bottome of boyled Artichokes, minced Marrow over and in the bottom of the Pye after your Pye is baked; when you put it up, have some five yolks of Eggs minced, and the juyce of two or three Oranges, the meat of one Lemon cut in pieces, a little White and Claret Wine; put this in your Pye being well mingled, and shake it very well together.

To boyle a Capon or Hen.

Take a young Capon or Hen, when you draw them, take out the fall of the Leafe clean away, and being well washed, fill the belly with Oysters; prepare some Mutton, the neck, but boyle it in smal peices and skim it well, then put your Capon into the Pipkin, and when it is boyled, skim it again; be sure you have no more water then will cover your meat, then put it into a pint of white wine, some Mace, two or three Cloves and whole Pepper; a quarter of an hour before your meat be boyled enough, put into the Pipkin, three Anchoves stript from the Bones and washed, and be sure you put Salt at the first to your meat; a little Parsley Spinage, Endive, Sorrell,

Rose-mary, or such kind of Herbs will do well to boyle with the Broth, and being ready to Dish it, having sippets cut then take the Oysters out of the Capon, and lay them in the Dish with the Broth, and put some juyce of Lemons and Orange into it according to your taste.

To make Balls of Veale.

Take the Lean of a Leg of Veal, and cut out the Sinews, mince it very small, and with it some fat of Beef suet; if the Leg of Veal be of a Cow Calfe, the Udder will be good instead of Beef suet; when it is very well beaten together with the mincing Knife, have some Cloves, Mace, and Pepper beaten, and with Salt season your meat, putting in some Vinegar, then make up your meat into little Balls, and having very good strong Broth made of Mutton, set your Balls to boyle in it; when they are boyled enough, take the yolks of five or six Eggs well beaten with as much Vinegar as you please to like, and some of the Broth mingled together, stir it into all your Balls and Broth, give it a waume on the fire, then Dish up the Balls upon Sippits and pour the sauce on it.

To make Mrs. Shellyes Cake.

Take a peck of fine flower, and three pound of the best Butter, work your flower and butter very well together, then take ten Eggs, leave out six whites, a pint and a halfe of Ale-yeast: beat the Eggs and yeast together, and put them to the flower; take six pound of blanched Almonds, beat them very well, putting in sometime Rosewater to keepe them from Oyling; adde what spice you please; let this be put to the rest, with a quarter of a pint of Sack, and a little saffron; and when you have made all this into Past, cover it warme before the fire, and let it rise for halfe an hour, then put in twelve pound of Currans well washed and dryed, two pound of Raisins of the Sun stoned and cut small, one pound of Sugar; the sooner you put it into the Oven after the fruit is put in, the better.

To make Almond Jumballs.

Take a pound of Almonds to halfe a pound of double refined Sugar beaten and Searced, lay your Almonds in water a day before you blanch them, and beat them small with your Sugar; and when it is beat very small, put in a handfull of Gum-dragon, it being before over night steeped in Rose-water, and halfe a white of an Egge beaten to froth, and halfe a spoonfull of Coriander-seed as many Fennell and Ani-seeds, mingle these together very well, set them upon a soft fire till it grow pretty thick, then take it off the fire, and lay it upon a clean Paper, and beat it well with a rowling pin till it work like a soft past, and so make them up, and lay them upon Papers oyld with Oyle of Almonds, then put them in your Oven, and so soon as they be throughly risen, take them out before they grow hard.

To make Cracknels.

Take halfe a pound of fine flower, dryed and searced, as much fine sugar searced, mingled with a spoonfull of Coriander-seed bruised, halfe a quarter of a pound of butter rubbed in the flower and sugar, then wet it with the yolks of two Eggs, and halfe a spoonfull of white Rose-water, a spoonfull or little more of Cream as will wet it; knead the Past till it be soft and limber to rowle well, then rowle it extreame thin, and cut them round by little plates; lay them up on buttered papers, and when they goe into the Oven, prick them, and wash the Top with the yolk of an Egg beaten, and made thin with Rose-water or faire water; they will give with keeping, therefore before they are eaten, they must be dryed in a warme Oven to make them crisp.

To Pickle Oysters.

Take Oysters and wash them cleane in their own Liquor, and let them settle, then strain it, and put your Oysters to it with a little Mace and whole pepper, as much Salt as you please, and a little Wine-Vinegar, then set them over the fire, and let them boyle leisurely till they are pretty tender; be sure to skim them still as the skim riseth; when they are enough, take them out till the Pickle be cold, then put them into any pot that will lye close, they will keep best in Caper barrels, they will keep very well six weeks.

To boyle Cream with Codlings.

Take a quart of Cream and boyle it with some Mace and Sugar, and take two yolks of Eggs, and beat them well with a spoonfull of Rose-water and a grain of Amber-greece, then put it into the Cream with a piece of sweet Butter as big as a Wall-nut, and stir it together over the fire untill it be ready to boyle, then set it some time to coole, stirring it continually till it be cold; then take a quarter of a pound of Codlings strained, and put them into a silver Dish over a few coales till they be almost dry, and being cold, and the Cream also, poure the Cream upon them, and let them stand on a soft fire covered an hour, then serve them in.

To make the Lady Albergaveres Cheese.

To one Cheese take a Gallon of new Milk, and a pint of good Cream, and mix them well together, then take a Skillet of hot water as much as will make it hotter then it comes from the Cow, then put in a spoonfull of Rennet, and stir it well together and cover it, and when it is come, take a wet Cloth and lay it on your Cheese-Mot, and take up the Curd and not break it; and put it into your Mot; and when your Mot is full, lay on the Suiker, and every two hours turn your Cheese in wet Cloathes wrung dry; and lay on a little more wet, at night take as much salt as you can between your finger and thumb, and salt your Cheese on both sides; let them lye in Presses all night in a wet Cloth; the next day lay them on a Table between a dry Cloth, the next day lay them in Grasse, and every other day change your Grasse, they will be ready to eat in nine dayes; if you will have them ready sooner, cover them with a Blanket.

To dresse Snayles.

Take your Snayles (they are no way so as in Pottage) and wash them well in many waters, and when you have done put them in a white Earthen Pan, or a very wide Dish, and put as much water to them as will cover them, and then set your Dish or Pan on some coales, that it may heat by little and little, and then the Snayles will come out of the shells and so dye, and being dead, take them out, and wash them very well in Water and salt twice or thrice over; then put them in a Pipkin with Water and Salt, and let them boyle a little while in that, so take away the rude slime they have, then take them out againe and put them in a Cullender; then take excellent sallet Oyle and beat it a great while upon the fire in a frying Pan, and when it boyls very fast, slice two or three Onyons in it, and let them fry well, then put the Snayles in the Oyle and Onyons, and let them stew together a little, then put the Oyle, Onyons, and Snayles altogether in an earthen Pipkin of a fit size for your Snayles, and put as much warm water to them as will serve to boyle them, and make the Pottage and season them with Salt, and so let them boyle three or foure hours; then mingle Parsly, Pennyroyall, Fennell, Tyme, and such Herbs, and when they are minced put them in a Morter, and beat them as you doe for Green-sauce, and put in some crums of bread soaked in the Pottage of the Snayles, and then dissolve it all in the Morter with a little Saffron and Cloves well beaten, and put in as much Pottage into the Morter as will make the Spice and bread and Herbs like thickning for a pot, so put them all into the Snayles and let them stew in it, and when you serve them up, you may squeeze into the pottage a Lemon, and put in a little Vinegar, or if you put in a Clove of Garlick among the Herbs, and beat it with them in the Morter; it will not tast the worse; serve them up in a Dish with sippets of Bread in the bottom. The Pottage is very nourishing, and they use them that are apt to a Consumption.

To boyle a rump of Beefe after the French fashion.

Take a rump of Beef, or the little end of the Brisket, and parboyle it halfe an houre, then take it up and put it in a deep Dish, then slash it in the side that the gravy may come out, then throw a little Pepper and salt betweene every cut, then fill up the Dish with the best Claret wine, and put to it three or four pieces of large Mace, and set it on the coales close covered, and boyle it above an houre and a halfe, but turn it often in the mean time; then with a spoon take of the fat and fill it with Claret wine, and slice six Onyons, and a handfull of Cappers or broom buds, halfe a dozen of hard Lettice sliced, three spoonfuls of wine-Vinegar and as much verjuyce, and then set it a boyling with these things in it till it be tender, and serve it up with brown Bread and Sippets fryed with butter, but be sure there be not too much fat in it when you serve it.

An excellent way of dressing Fish.

Take a piece of fresh Salmon, and wash it clean in a little Vinegar and Water, and let it lye a while in it, in a great Pipkin with a cover, and put to it six spoonfuls of Water and four of Vinegar, as much of white wine, a good deale of Salt, a bundle of sweet Herbs, a little whole Spice, a few Cloves, a little stick of Cinamon, a little Mace, take up all these in a Pipkin close, and set in a Kettle of seething Water and there let it stew three hours, You may doe Carps, Eeles, Trouts, &c. this way, alter the tast to your mind.

To make Fritters of Sheeps-feet.

Take your Sheeps feet, slit them and set them a stewing in a silver Dish, with a little strong Broth and Salt, with a stick of Cinamon, two or three Cloves, and a piece of an Orange Pill; when they are stewed, take them from the liquor and lay them upon a Pye-plate cooling; when they are cold, have some good Fritter-batter made with Sack, and dip them therein; then have ready to fry them, some excellent clarified Butter very hot in a Pan, and fry them therein; when they are fryed wring in the juyce of three or four Oranges, and toss them once or twice in a Dish, and so serve them to the Table.

To make dry Salmon Calvert in the boyling.

Take a Gallon of Water, put to it a quart of Wine or Vinegar, Verjuyce or sour Beer, and a few sweet herbs and Salt, and let your Liquor boyle extream fast, and hold your Salmon by the Tayle, and dip it in, and let it have a walme, and so dip it in and out a dozen times, and that will make your Salmon Calvert, and so boyle it till it be tender.

To make Bisket Bread.

Take a pound of Sugar searced very fine, and a pound of flower well dryed, and twelve Eggs, a handfull of Carroway-seed, six whites of Eggs, a very little Salt, beat all these together, and keep them with beating till you set them in the Oven, then put them into your Plates or Tin things, and take Butter and put into a Cloth and rub your Plate; a spoonfull into a Plate is enough, and so set them in the Oven, and let your Oven be no hotter then to bake small Pyes; if your flower be not dryed in the Oven before, they will be heavy.

To make an Almond Pudding.

Take your Almonds when they are blanched, and beat them as many as will serve for your Dish, then put to it foure or five yolks of Eggs, Rose-water, Nutmeg, Cloves and Mace, a little Sugar, and a little salt and Marrow cut into it, and so set it into the Oven, but your Oven must not be hotter then for Bisket bread; and when it is half baked, take the white of an Egg, Rose-water and fine Sugar well beaten together and very thick, and do it over with a feather, and set it in againe, then stick it over with Almonds, and so send it up.

This you may boyle in a Bag if you please, and put in a few crums of Bread into it, and eat it with butter and Sugar without Marrow.

To make an Almond Caudle.

Take three pints of Ale, boyle it with Cloves and Mace, and sliced bread in it, then have ready beaten a pound of Almonds blanched, & strain them out with a pint of White wine, and thicken the Ale with it, sweeten it if you please, and be sure you skim the Ale well when it boyles.

To make Almond bread.

Take Almonds and lay them in water all night, then blanch them and slice them, to every pound of Almonds, a pound of fine Sugar finely beaten, so mingle them together, then beat the whites of three Egs to high froth, and mix it well with the Almonds & Sugar, then have some Plates and strew some flower on them, and lay Wafers on them, and lay your Almonds with the edges upwards, lay them as round as your can, scrape a little Sugar on them, when they are ready to set in the Oven, which must not be so hot as to colour white Paper, and when they are a little baked, take them out, and them from the Plates, and set them in again, you must keepe them in a Stove.

To make Almond Cakes.

Take halfe a pound of Almonds blanched in cold water, beat them with some Rose-water till they doe not glister, then they will be beaten; if you think fit, lay seven or eight Musque Comfits dissolved in Rosewater which must not be above six or seven spoonfuls for fear of spoyling the colour; when they be thus beaten, put in half a pound of Sugar finely sifted, beat them and the Almonds together till it be well mixed, then take the whites of two Eggs, and two spoonfuls of fine flower that hath been dried in an Oven; beat these wel together and poure it to your Almonds, then butter your Plates and dust your Cakes with Sugar and Flower, and when they are a little brown, draw them, and when the oven is colder set them in again on browne Papers, and they will looke whiter.

Master Rudstones *Posset.*

Take a Pint of Sack, a quarter of a pint of Ale three quarters of a pound of Sugar, boyle all these well together, take two yolks of Eggs and sixteen whites very well beaten, put this to your boyling Sack & slice it very well together till it be thick on the coales; then take the three pints of Milk or Cream being boyled to a quart, it must stand and cool till the Eggs thicken, put it to your Sack and

Eggs, and stir them well together, then cover it with a Plate and so serve it.

To boyle a Capon with Ranioles.

Take a good young Capon, trusse it and draw it to boyle, and parboyle it a little, then let it lye in fair Water being pickt very cleane and white, then boyle it in strong broth while it be enough, but first prepare your Ranioles as followeth; Take a good quantity of Beet leaves, and boyle them in Water very tender, then take them out, and get all the water very cleane out of them, then take six sweet breads of Veale, and boyle and mince them white, mince them small, and then boyl Herbs also, and four or five Marrow bones, break them and get all the Marrow out of them, and boyle the bigger peice of them in water by it selfe, and put the other into the minced Herbs, then take halfe a pound of Raisins of the Sun stoned, and mince them small, and halfe a pound of Dates the skin off, and mince them also, and a quarter of a pound of Pomecitron minced small, then take of Naples-bisket a good quantity, and put all these together on a Charger or a great Dish with halfe a pound of sweet Butter, and worke it together with your hands as you do a peice of Past, and season it with a little Nutmeg, Ginger, Cinamon, and Salt, & Permasan Cheese grated with hard Sugar grated also, then mingle all together well, and make a Past with the finest flower, six yolks of Eggs, a little Saffron beaten small, halfe a pound of sweet Butter, a little Salt, with some faire water hot (not boyling) and make up your Past, then drive out a long sheet of Past with an even Rowling Pin as thin as possible you can, and lay your ingredients in small heaps, round or long which you please in the Past, then cover them with the Past & cut them with a jag asunder and so make more or more till you have made two hundred or more, then have a good broad Pan or Kettle halfe full of strong Broth, boyling leisurely, and put in your Ransoles one by one, and let them boyle a quarter of an hour, then take up your Capon, lay it in a great Dish, and put one the Ransoles, & strew on them grated Cheese, Naples-Bisket grated, Cinamon and Sugar, then more and more Cinamon & Cheese, while you have filled your Dish; then put softly on melted Butter with a little strong Broth, your Marrow Pomecitron, Lemons sliced and serve it up, and so put it into the Dish so Ransoles may be part fryed with sweet but Clarified butter, either a quarter of them or halfe as you please; if the butter be not Clarified, it will spoile your Ransoles.

To make a Bisque of Carps.

Take twelve small Carps, and one great one, all Male Carps, draw them and take out all the Melts, flea the twelve small Carps, cut off their Heads and take out their Tongues and take the fish from the bones of the flead Carps, and twelve Oysters, two or three yelks of Hard eggs, mash altogether, season it with Cloves, Mace and Salt, and make thereof a stiffe searce, add thereto the yolks of foure or five Eggs to bind it, fashion that first into bals or Lopings as you please, lay them into a deep Dish or Earthen Pan, and put thereto twenty or thirty great Oysters, two or three Anchoves, the Milts and Tongues of your twelve Carps, halfe a pound of fresh butter, the Liquor of your Oysters, the juyce of a Lemon or two; a little Whitewine some of Corbilion wherein your great Carpe is boyled, and a whole Onyon, so set them a stewing on a soft fire and make a hoop therewith; for the great Carp you must scald him and draw him, and lay him for half an hour with the other Carps Heads in a deep Pan with so much White wine Vinegar as will cover and serve to boyle him, and the other Heads in; put therein Pepper, whole Mace, a race of Ginger, Nutmeg, Salt sweet Herbs, an Onyon or two sliced, a Lemon; when you boyle your Carps, poure your Liquor with the Spice into the Kettle wherein you will boyle him; when it is boyled put in your Carp, let it not boyle too fast for breaking; after the Carp hath boyled a while put in the Head, when it is enough take off the Kettle, and let the Carps and the Heads keep warme in the Liquor till you goe to dish them. When you dresse your Bisque, take a large Silver dish, set it on the fire, lay therein Sippets of bread, then put in a Ladle-full of your Corbilion, then take up your great Carp and lay him in the midst of the Dish, then range the twelve heads about the Carp, then lay the searce of the Carp, lay that in, then your Oysters, Milts, and Tongues, then poure on the Liquor wherein the searce was boyled, wring in the juyce of a Lemon and two Oranges; Garnish your Dish with pickled Barberries, Lemons and Oranges, and serve it very hot to the Table.

To boyle a Pike and Eele together.

Take a quart of White-Wine and a pint and a halfe of White-Wine-Vinegar, two quarts of Water, and almost a pint of Salt, a handfull of Rose-mary and Tyme; the Liquor must boyle before you put in your Fish and Herbs; the Eele with the skins must be put in a quarter of an hour before the Pike, with a little large Mace, and twenty cornes of Pepper.

To make an Outlandish dish.

Take the liver of a Hogg, and cut it in small pieces about the bigness of a span, then take Anni-seed, or French-seed, Pepper and Salt, and season them therewithall, and lay every piece severally round in the caule of the Hogg, and so roast them on a Bird-Spit.

To make a Portugall Dish.

Take the Guts, Gizards and Liver of two fat Capons, cut away the Galles from the Liver, and make clean the Gizards and put them into a Dish of clean water, slit the Gut as you do a Calves Chaldron but take off none of the fat, then lay the Guts about an hour in White-wine, as the Guts soke, half boyle Gizards and Livers, then take a long wooden broach, and spit your Gizards and Liver thereon, but not close one to another, then take and wipe the Guts somewhat dry in Cloth, and season them with Salt and beaten Pepper, Cloves and Mace, then wind the Guts upon the wooden Broach about the Liver, and Gizards, then tye the wooden Broach to spin, and lay them to the fire to roast, and roast them very brown, and bast them not at all till they be enough, then take the Gravy of Mutton, the juyce of two or three Oranges, and a grain of Saffron, mix all well together, and with a spoon bast your roast, let it drop into the same Dish. Then draw it, and serve it to the Table with the same sauce.

To dresse a dish of Hartichoaks.

Take and boyle them in the Beef-pot, when they are tender sodden, take off the tops, leaving the bottoms with some round about them, then put them into a Dish, put some fair water to them, two or three spoonfuls of Sack, a spoonfull of Sugar, and so let them boyle upon the Coales, still pouring on the Liquor to give it a good tast, when they have boyled halfe an hour take the Liquor from them, and make ready some Cream boyled and thickned with the yolk of an Egge or two, whole Mace, Salt, and Sugar with some lumps of marrow, boyle it in the Cream, when it is boyled put a good piece of sweet butter into it, and toast some toasts, and lay them under your Hartichoaks, and poure your Cream, and butter on them, Garnish it, &c.

To dresse a Fillet of Veale the Italian way. Take a young tender Fillet of Veale, pick away all the skins in the fold of the flesh, after you have picked it out clean, so that no skins are left, nor any hard thing; put to it some good White-Wine (that is not too sweet) in a bowl & wash it, & crush it well in the Wine; doe so twice, then strew upon it a powder that is called *Tamara* in *Italy*, and so much Salt as will season it well, mingle the Powder well upon the Pasts of your meat, then poure to it so much White-Wine as will cover it when it is thrust down into a narrow Pan; lay a Trencher on it and a weight to keep it downe, let it lye two nights and one day, put a little Pepper to it when you lay it in the Sauce, and after it it is sowsed so long, take it out and put it into a Pipkin with some good Beef-broth, but you must not take any of the pickle to it, but onely Beef-broth that is sweet and not salt; cover it close and set it on the Embers, onely put into it with the Broth a few whole Cloves and Mace, and let it stew till it be enough. It will be very tender and of an excellent Taste; it must be served with the same broth as much as will cover it.

To make the Italian, take Coriander seed two Ounces, Aniseed one ounce, Fennel-seed one ounce, Cloves two ounces, Cinamon one ounce; These must be beaten into a grosse powder, putting into it a little powder of Winter-savoury; if you like it, keep this in a Vial-glasse close stopt for your use.

To dresse Soales.

Take a Pair of Soales, lard them through with watered fresh Salmon, then lay your Soales on a Table, or Pie-plate, cut your Salmon, lard all of an equal length on each side, and leave the Lard but short, then flower the Soales, and fry them in the best Ale you can get; when they are fryed lay them in a warme Pie-plate, and so serve them to the Table with a Sallet dish full of Anchovy sauce, and three or four Oranges.

To make Furmity.

Take a quart of Creame, a quarter of a pound of French-barley the whitest you can get, and boyle it very tender in three or four severall waters, and let it be cold, then put both together, put in it a blade of Mace, a Nutmeg cut in quarters, a race of Ginger cut in three or five pieces, and so let it boyle a good while, still stirring, and season it with Sugar to your tast, then take the yolks of four Eggs and beat them with a little Cream, and stir them into it, and so let it boyle a little after the Egs are in, then have ready blanched and beaten twenty Almonds kept from oyling, with a little Rose-water, then take a boulter, strainer, and rub your Almonds with a little of your Furmity through the strainer, but set on the fire no more, and stir in a little Salt and a little sliced Nutmeg, pickt out of the great peices of it, and put it in a dish, and serve it.

To make Patis, or Cabbage Cream.

Take thirty Ale pints of new milke, and set it on the fire in a Kettle till it be scalding hot, stirring it oft to keep it from creaming, then put in forth, into thirty Pans of Earth, as you put it forth, take off the bubbles with a spoon, let it stand till it be cold, then take off the Cream with two such slices as you beat Bisket bread with, but they must be very thin and not too broad, then when the Milk is dropped off the Cream, you must lay it upon a Pye-plate, you must scour the Kettle very clean and heat the Milk again, and so four or five times. In the lay of it, first lay a stalk in the midst of the Plate, let the rest of the Cream be laid upon that sloping, between every laying you must scrape Sugar and sprinkle Rose-water, and if you will, the powder of Musk, and Amber-greece, in the heating of the Milk be carefull of smoak.

To make Pap.

Take three quarts of new milk, set it on the fire in a dry silver Dish, or Bason, when it begins to boyle skim it, then put thereto a handfull of flour & yolks of three Eggs, which you must have well mingled together with a Ladle-full of cold Milk, before you put it to the Milk that boyles, and as it boyles, stir it all the while till it be enough, and in the boyling, season it with a little Salt, and a little fine beaten Sugar and so keeping it stirred till it be boyled as thick as you desire, then put it forth into another Dish and serve it up.

To make Spanish Pap.

Take three spoonfuls of Rice-floure, finely beaten and searced, two yolks of Eggs, three spoonfuls of Sugar, three or foure spoonfuls of Rose-water. Temper these fouer together, then put them to a pint of cold Cream, then set it on the fire and keep it stirred till it come to a reasonable thicknesse, then Dish it and serve it up.

To poach eggs.

Take a dozen of new laid Eggs and flesh of foure or five Partridges, or other; mince it so small as you can season it with a few beaten Cloves, Mace, and Nutmeg, into a Silver Dish, with a Ladlefull or two of the Gravy of Mutton, wherein two or three Anchoves are dissolved; then set it a stewing on a fire of Char-Coales, and after it is halfe stewed, as it boyles, break in your Eggs, one by one, and as you break them, poure away most part of the Whites, and with one end of your Egg-shell, make a place in your Dish of meat, and therein put your Yolks of your Eggs, round in order amongst your meat, and so let them stew till your Eggs be enough, then grate in a little Nutmeg, and the juyce of a couple of Oranges; have a care none of the Seeds goe in, wipe your Dish and garnish your Dish, with four or five whole Onions,&c.

A Pottage of Beef Pallats.

Take Beefe Pallats after they be boyled tender in the Beefe Kettle, or Pot among some other meat, blanch and serve them cleane, then cut each Pallat in two, and set them a stewing between two Dishes with a piece of leer Bacon, an handful of Champignions, five or six sweetbreads of Veale, a Ladle-full or two of strong broth, and as much gravy of Mutton, an Onion or two, five or six Cloves, and a blade or two of Mace, and a piece of Orange Pils; as your Pallats stew, make ready your Dish with the bottoms and tops of two or three Cheat Loaves dryed and moystned with some Gravy of Mutton, and the broth your Palats stew in, you must have the Marrow of two or three beef-bones stewed in a little broth between two Dishes in great pieces; when your Pallats and Marrow iss stewed, and you ready to Dish it, take out all the Spices, Onyon and Bacon, and lay it in your Plates, sweetbread, and Champigneons, pour in the Broath they were stewed in & lay on your peices of Marrow, wring the juyce of two or three Oranges; and so serve it to the Table very hot.

The Jacobins Pottage.

Take the flesh of a washed Capon or Turkey cold, mince it so small as you can, then grate or scrape among the flesh two or three ounces of Parmasants or old Holland Cheese, season it with beaten Cloves, Nutmeg, Mace, and Salt, then take the bottoms and tops of foure or five new Rowles, dry them before the fire, or in an Oven, then put them into a faire silver Dish set it upon the fire, wet your bread in a Ladle full of strong Broth, and a Ladle-full of Gravy of Mutton then strow on your minced meat all of an equall thicknesse in each place, then stick twelve or eighteen peices of Marrow as bigge as Walnuts, and pour on an handfull of pure Gravy of Mutton then cover your Dish close, and as it stews adde now and then some Gravy of Mutton there to, thrust your Knife sometimes to the bottome, to keep the bread from sticking to the Dish, let it so stew stil, till you are ready to Dish it away, and when you serve it, if need require, ad more Gravy of Mutton, wring the juyce of two or three Oranges, wipe your Dishes brims, and serve it to the Table in the same Dish.

To Salt a Goose.

Take a fat Goose and bone him, but leave the brest bone, wipe him with a clean cloath, then salt him one fortnight, then hang him up for one fortnight or three weeks, then boyl him in running water very tender, and serve him with Bay-leaves.

A way of stewing Chickens or Rabbets.

Take two three or foure Chickens, and let them be about the bigness of a Partridge, boyl them til they be half boyled enough, then take them off and cut them into little peices, putting the joynt bone one from another, and let not the meat be minced, but cut into great bits, not so exactly but more or lesse, the brest bones are not so proper to be put in, but put the meat together with the other bones (upon which there must also be some meat remaining) into a good quantity of that Water or Broth wherein the Chickens were boyled, and set it then over a Chaffing-Dish of coales betweeen two Dishes, that so it may stew on till it be fully enough; but first season it with Salt and gross Pepper, and afterwards add Oyl to it, more or lesse according to the goodnesse thereof; and a little before you take it from the fire, you must adde such a quantity of juyce of Lemons as may best agree with your Taste. This makes an excellent dish of Meat, which must be served up in the Liquor; and though for a need it may be made with Butter instead of Oyl, and with Vinegar in stead of Juyce of Lemons, yet is the other incomparably better for such as are not Enemies to Oyle. The same Dish may be made also of Veal, or Partridge, or Rabbets, and indeed the best of them all, is Rabbets, if they be used so before Michaelmas, for afterwards me-thinkes they grow ranke; for though they be fatter, yet the flesh is more hard and dry.

A Pottage of Capons.

Take a couple of young Capons, Trusse and set them and fill their bellies with Marrow, put them into a Pipkin with a knuckle of Veale, a Neck of Mutton, and a Marrow bone, and some sweet bread of Veale; season your Broth with Cloves, Mace, and a little Salt, set it to the fire, and let it boyle gently till your Capons be enough, but boyle them not too much; as your Capons boyle, make ready the bottomes and Tops of eight or ten new Rowles, and put them dryed into a faire Silver Dish wherein you serve the Capons; set it on the fire, and put to your bread, two Ladlefuls of Broth wherein your Capons are boyled and a Ladlefull of the Gravy of Mutton; so cover your Dish, and let it stand till you Dish up yovr Capons if need require, adde now and then a Ladlefull of Broth and Gravy, least the bread grow dry; when you are ready to serve it, first lay in the Marrow bone, then the Capons on each side, then fill up your Dish with the Gravy of Mutton, wherein you must wring the juyce of a Lemon or two, then with a spoon take off all the fat that swimmeth on the pottage, then garnish your Capon with the sweet Breads and some Lemons, and so serve it.

To dresse Soales another way.

Take Soales, fry them halfe enough, then take Wine seasoned with Salt, grated Ginger, and a little Garlick, let the Wine, and seasoning boyle in a Dish, when that boyles and your Soales are halfe fry'd, take the Soales and put them into the Wine, when they are sufficiently stewed, upon their backs, lay the two halfs open on the one side and on the other, then lay Anchoves finely washed along, and on the sides over again, let them stew till they be ready to be eaten, then take them out, lay them on the Dish, pour some of the clear Liquor which they stew in upon them, and squeeze an Orange in.

A Carpe Pye.

Take Carps scald them, take out the great bones, pound the Carps in a stone Morter pound some of the blood with the flesh which must be at the discretion of the Cook because it must not be too soft, then lard it with the belly of a very fat Eale, season it, and bake it like red Deere and eat it cold.

This is meat for a Pope.

To boyle Ducks after the french fashion.

Take and lard them and put them upon a spit, and halfe roast them, then draw them & put them into a Pipkin, and put a quart of Clarit Wine into it, and Chesnuts, & a pint of great Oysters taking the beards from them, and three Onyons minced very small, some Mace and a little beaten Ginger, a little Tyme stript, a Crust of a French Rowle grated put into it to thicken it, and so dish it upon sops. This may be diversified, if there be strong broth there need not be so much Wine put in, and if there be no oysters or Chesnuts you may put in Hartichoak bottoms, Turnips, Colliflowers, Bacon in thin slices, Sweet bread's, &c.

To boyle a Goose with Sausages.

Take your Goose and salt it two or three dayes, then trusse it to boyle, cut Lard as big as the top of your finger, as much as will Lard the flesh of the brest, season your lard with Pepper, Mace, and Salt; put it a boyling in Beefe broth if you have any, or water, season your Liquor with a little Salt, and Pepper grosly beaten an ounce or two, a bundle of Bay-leaves, Rosemary and Tyme, tyed altogether; you must have prepared your Cabbage or sausages boyl'd very tender, squeese all the water from them, then put them into a Pipkin, put to them a little strong broth or Claret Wine, an Onyon or two; season it with Pepper, Salt and Mace to your tast; six Anchoves dissolved, put altogether, and let them stew a good while on the fire; put a Ladle of thicke Butter, a little Vinegar, when your Goose is boyled enough, and your Cabbage on Sippets of bread and the Goose on the top of your Cabbage, and some on the Cabbage on top of your Goose, serve it up.

To fry Chickens.

Take five or six and scald them, and cut them in pieces, then flea the skin from them, fry them in Butter very brown, then take them out, and put them between two Dishes with the Gravy of Mutton, Butter, and an Onyon, six Anchoves, Nutmeg, and salt to your taste, then put sops on your Dish, put fryed Parsley on the top of your Chicken being Dished, and so serve them.

To make a Battalia Pye.

Take four tame Pigeons and Trusse them to bake, and take foure Oxe Pallats well boyled and blanched, and cut it in little pieces; take six Lamb stones, and as many good Sweet breads of Veale cut in halfs and parboyl'd, and twenty Cockscombs boyled add blanched, and the bottoms of four Hartichoaks, and a Pint of Oysters parboyled and bearded, and the Marrow of three bones, so season all with Mace, Nutmeg and Salt; so put your meat in a Coffin of Fine Paste proportionable to your quantity of meat; put halfe a pound of Butter upon your meat, put a little water in the Pye, before it be set in the Oven, let it stand in the Oven an houre and a halfe, then take it out, pour out the butter at the top of the Pye, and put it in leer of Gravy, butter, and Lemons, and serve it up.

To make a Chicken Pye.

Take four or five chickens, cut them in peices, take two or three Sweet-breads parboyl'd and cut the peices as big as walnuts; take the Udder of Veal cut in thin slices, or little slices of Bacon, the bottom of Hartichoaks boyl'd, then make your coffin proportionable to your meat, season your meat with Nutmeg, Mace and Salt, then some butter on the top of the Pye, put a little water into it as you put it into the Oven, and let it bake an hour, then put in a leer of butter, Gravy of Mutton, eight Lemons sliced; so serve it.

To make a Pye of a Calves head.

Take a Calves head, cleane it and wash it very well, put it a boyling till it be three quarters boyled, then cut off the flesh from the bones, and cut it in peices as big as Walnuts. Blaunch the Tongue and cut it in slices, take a quart of Oysters parboil'd and bearded, take the yolks of twelve Eggs, put some thin slices of bacon among the meat, and on the top of the meat, when it is in the Pye cut an Onion small, and put it in the bottom of your Pye, season it with Pepper, Nutmeg, Mace, and Salt, make your Coffin to your meat what fashion you please. Let it bake an hour and a half, put butter on the bottom and on the top of your Pye before you close it, put a little water in before you put it into the Oven, when you draw it out take off the Lid, and put away all the fat on the top and put in a leer of thick butter, Gravy of Mutton, a Lemon pared and sliced with two or three Anchoves dissolved. So stew these together, and cut your Lid in handsome peices, and lay it round the Pye, so serve it.

To make Creame with Snow.

Take three Pints of Creame, and the whites of seven or eight Eggs and strain them together, and a little Rose-water, and as much Sugar as will sweeten it, then take a sticke as big as a childs Arme, cleave one end of it a crosse, and widen your peices with your finger, beat your Cream with this sticke, or else with a bundle of Reeds tyed together, and rowl between your hand standing upright in your Creame, now as the Snow ariseth take it up with a spoon in a Cullender that the thin may run out, and when you have sufficient of this Snow; take the Cream that is left, & seeth it in the Skellet, and put thereto whole Cloves, stickes of Cinnamon, a little Ginger bruised, and seeth it till it be thick, then strain it, and when it is cold put it into your Dish, and lay your Snow upon it.

To make minced Pies.

Take a large Neats tongue, shread it very well, three pound and a halfe of Suet very well shread, Currans three pound, halfe an ounce of beaten Cloves and Mace, season it with Salt when you think't fit, halfe a preserved Orange, or instead of it Orange Pils, a quarter of a pound of Sugar, and a little Lemon Pill sliced very thin, put all these together very well, put to it two Spoonfull of Verjuyce, and a quarter of a Pint of Sack, &c.

To dry Neats Tongues.

Take Bay salt beaten very fine, and Salt-Peeter of each a like, and rub your Tongues very well with that, and cover all over with it, and as it wasts put on more, and when they are very hard and stiffe they are enough, then rowle them in Bran, and dry them before a soft fire, and before you boyle them, let them lie one night in Pompe Water, and boyle them in the same sort of water.

To make Jelly of Harts Horn.

Take six ounces of Hart-Horn, three ounces of Ivory both finely carped, boyle it in two quarts of water in a Pipkin close covered, and when it is three parts wasted, you may try it with a Spoon if it will be jelly, you may know by the sticking to your Lips, then straine it through a jelly bag; season it with Rose-water, juyce of Lemons and double refined Sugar, each according to your Taste, then boyle altogether two or three walmes, so put in the Glasse and keep for your use.

To make Chickens fat in four or five dayes.

Take a pint of French Wheat and a pint of Wheat flower, halfe a pound of Sugar, make it up into a stiff Paste, and rowle it into little rowles, wet them in warme Milk, and so Cram them, and they will be fat in four or five dayes, if you please you may sow them up behind one or two of the last dayes.

To make Angelot.

Take a Gallon of Stroakings and a Pint of Creame as it comes from the Cow, and put it together with a little Rennet; when you fill, turne up the midst side of the Cheese-fat, fill them a little at once, and let it stand all that day and the next, then turn them, and let them stand til they will slip out of the Fat, Salt them on both sides, and when the Coats begin to come on them, neither wipe nor scrape them, for the thicker the Coat is the better.

A Persian Dish.

Take the fleshly part of a Leg of Mutton stript from the fat and sinews, beat that well in a Morter with Pepper and Salt, and a little Onyon or Garlick water by it selfe, or with Herbs according to your taste, then make it up in flat cakes and let them be kept twelve houres betweene two Dishes before you use them, then fry them with butter in a frying Pan and serve them with the same butter, and you will find it a dish of savory meat.

To roast a shoulder of Mutton in blood.

When your sheepe is killed save the blood, and spread the caule all open upon a Table that is wet, that it may not stick to it; as soone as you have flead your sheepe, cut off a shoulder, and having Tyme picked, shred and cut small into some of your blood, stop your shoulder with it, inside and outside, and into every hole with a Spoone, put some of the Blood; after you have put in the Tyme, then lay your Shoulder of Mutton upon the caule and wrap that about it, then lay it into a Tray, and pour all the rest of the blood upon it; so let it lie all night, if it be in Winter, you may let it lie twenty foure hours, then roast it.

To roast a Leg of Mutton to be eaten cold.

First take so much Lard as you thinke sufficient to Lard your Leg of Mutton withall, cut your Lard in grosse long Lardors; season the Lard very deep with beaten Cloves, Pepper, Nutmeg, and Mace, and bay salt beaten fine and dryed, then take Parsley, Tyme, Marjoram, Onion, and the out-rine of an Orange, shred all these very small, and mix them with the Lard, then Lard your Legge of Mutton therewith, if any of the Herbs and Spice remaine, put them on the Legge of Mutton; then take a silver Dish, lay two stickes crosse the Dish to keepe the Mutton from sopping in the Gravy and fat that goes from it, lay the Legge of Mutton upon the stickes, and set it into an hot Oven, there let it roast, turne it once but baste it not at all, when it is enough and very tender, take it forth but serve it not till it be throughly cold; when you serve it, put in a saucer or two of

Mustard, and Sugar, and two or three Lemons whole in the same dish.

To roast Oysters.

Take the greatest Oysters you can get, and as you open them, put them into a Dish with their own Liquor, then take them out of the Dish, and put them into another, and pour the Liquor to them, but be sure no gravell get amongst them; then set them covered on the fire, and scald them a little in their owne Liquor, and when they are cold, draw eight or ten Lards through each Oyster; season your Lard first with Cloves, Nutmeg beaten very small, Pepper; then take two woodden Lard Spits, and spit your Oysters thereon, then tye them to another spit, and roast them. In the roasting bast them with Anchovy sauce, made with some of the Oyster Liquor, and let them drip into the same dish where the Anchovy sauce is; when they be enough, bread them with the crust of a roul grated on them, and when they be brown, draw them off, then take the sauce wherewith you basted your Oysters, and blow off the fat, then put the same to the Oysters, wring in it the juyce of a Lemon, so serve it.

To make a Sack Posset.

Take a quart of Cream and boyle it very well with Sugar, Mace, and Nutmeg, take half a pint of Sack, and as much Ale, and boyle them well together with some Sugar, then put your Cream into your Bason to your Sacke, then heat a pewter dish very hot, and cover your Bason with it, and set it by the fire side, and let it stand there two or three houres before you eat it.

Another Sack Posset.

Take eight Eggs, yolks and whites, and beat them well together, straine them into a quart of Cream, season them with Nutmeg and Sugar, put to them a pint of Sack, stir them altogether, and put them into your Bason, and set them in the Oven no hotter then for a Custard, let it stand two hours.

To make a Sack Posset without Milk or Cream.

Take eighteen Eggs wites and all, taking out the treads, let them be beaten very well, take a pint of Sack and a quart of Ale boyled, and scum it, then put in three quarters of a pound of Sugar and a little Nutmeg, let it boyle a little together, then take it off the fire stirring the Eggs still, put into them two or three Ladle-fulls of drink, then mingle all together and set it on the fire, and keepe it stirring till you finde it thick, then serve it up.

To make a stump Pye.

Take a Leg of mutton, one pound and a half of the best Suet, mince both small together, then season it with a quarter of a pound of Sugar, and a small quantity of salt, and a little cloves & mace, then take a good handful of Parsly half as much Tyme, and mince them very small, and mingle them with the rest; then take six new laid Eggs and break them into the meat and worke it well together, and put it into the past; then upon the Top put Raisins, Currans and Dates a good quantity, cover and bake it, when it is baked, and when it is very hot, put into it a quarter of a Pint of White wine Vinegar, and strow Sugar upon it, and so serve it.

To make Mrs. Leeds *Cheese Cakes.*

Take six quarts of milk and ren it prety cold, and when it is tender come drayn from it your Whey in a strainer, then hang it up till all the Whey be dropt from it, then presse it, change it into dry cloaths till it wet the cloth no longer, then beat it in a stone Morter till it be like butter, then straine it through a thin strayner, mingle it with a pound and a halfe of butter with your hands, take one pound of Almonds, and heat them with Rosewater till they are like your Curd, then mingle them with the yolks of twenty Eggs and a quart of Cream, two great Nutmegs, one pound and a half of sugar, when your Coffins are ready and going to set in the Oven; then mingle them together, let your Oven be made hot enough for a Pigeon Pye, and let a stone stand up till the scorcthing be past, then set them in, half an hour will bake them well, your Coffins must be made with

Milk and Butter as stiffe as for other Past, then you must set them into a pretty hot Oven, and fill them full of Bran, and when they are harded, take them out, and with a Wing, brush out the Bran, they must be pricked.

To make Tarts called Taffaty Tarts.

First wet your Past with Butter and cold Water, and rowle it very thin, also then lay them in layes, and between every lay of Apples strew some Sugar, and some Lemon Pill, cut very small, if you please put some Fennell-seed to them; then put them into a stoak hot Oven, and let them stand an hour in or more, then take them out, and take Rose-water and Butter beaten together, and wash them over with the same, and strew fine Sugar upon them; then put them into the Oven again, let them stand a little while and take them out.

To make Fresh Cheese.

Take three pints of raw Cream and sweeten it well with Sugar, and set it over the fire, let it boyle a while, then put in some Damask-Rose-Water, keep it still stirring least it burn too, and when you see it thickned and turned, take it from the fire, and wash the strainer and Cheesefat with Rose-water, then rowl it too and fro in the Strainer to draine the Whey from the Curd, then take up the Curd with a spoon and put them into the Fat, let it stand till it be cold, then put it into the Cheese Dish with some of the Whey, and so serve it up.

To make Sugar Cakes or Jumbals.

Take two pound of flower, dry it and season it very fine, then take a pound of Loaf Sugar, and beat it very fine, and searce it, mingle your Flower and Sugar very well, then take a pound and a halfe of sweet Butter and wash out the Salt, and breake it into bits with your Flower and Sugar, then take yolks of foure new laid Eggs, and four or five spoonfuls of Sack, and four spoonfuls of Creame; beat

all these together, then put them into your Flower, and knead them to a Past, and make them into what fashion you please, and lay them upon Paper or Plates, and put them into the Oven, and be carefull of them, for a very little thing bakes them.
For Jumbals you must only adde the whites of two or three Eggs.

To hash a Shoulder of Mutton.

Take a Shoulder of Mutton and slice it very thin till you have almost nothing but the Bone, then put to the meat some Claret wine, a great Onion, some Gravy of Mutton, six Anchoves, a hand full of Capers, the tops of a little Tyme, mince them very well together, then take nine or tenne Egges, the juyce of one or two Lemons, to make it tart, and make leere of them, then put the meat all in a Frying-Pan over the fire till it be very hot; then put in the leere of Eggs and soak altogether over the fire till it be very thick; then boyle your bone, and put it on the top of your meat being Dished, Garnish your Dish with Lemons, serve it up.

To dresse Flounders or Playce with Garlick and Mustard.

Take Flounders very new, and cut all the Fins and Tailes, then take out the Guts and wipe them very clean, they must not be at all washt, then with your Knife scorch them on both sides very grosely; then take the Tops of Tyme and cut them very small, and take a little Salt, Mace, and Nutmeg, and mingle the Tyme and them together, and season the Flounders; then lay them on the Grid-iron and bast them with Oyle or Butter, let not the fire be too hot, when that side next the fire is brown; turn it, and when you turn it, bast it on both sides till you have broyl'd them brown, when they are enough make your sauce with Mustard two or three Spoonfull according to discretion, six Anchoves dissolved very well, about halfe a pound of butter drawn up with garlick, vinegar, or bruised garlick in other vinegar, rubb the bottome of your Dish with garlick. So put your sauce to them, and serve them, you may fry them if you please.

A Turkish Dish.

Take fat of Beefe or Mutton cut in thin slices, wash it well, put it into a pot that hath a close cover, then put into it a good quantity of clean pick'd rice, skim it very well; then put into it a quantity of whole Pepper, two or three whole Onyons; let all this boyle very well, then take out the Onyon and dish it in Sippets, the thicker it is the better.

To dresse a Pyke.

Cut him in peices, and strew upon him salt and scalding vinegar, boyle him in water and White wine, when he is boyling put in sweet Herbs, Onyon, Garlick, Ginger, Nutmeg, and salt; when he is boyled take him out of the Liquor, and let him drayn, in the mean time beat Butter and Anchoves together, and pour it on the fish, squeezing a little Orange and Lemon upon it.

To dresse Oysters.

Take Oysters and open them, and save the Liquor, and when you have opened so many as you please, adde to this Liquor, some White-wine, wherein you must wash your Oysters one by one very clean, and lay them in another Dish; then strain to them that mixed wine and Liquor wherein they were washed, adding a little more Wine to them with an Onion divided with some Salt and Pepper, so done, cover the Dish and stew them till they be more then halfe done; then take them and the Liquor, and pour it in to a Frying-Pan, wherein they must fry a pretty while, then put into them a good peice of sweet butter, and fry them therein so much longer; in the mean time you must have beaten the yolks of some Eggs, as four or five to a quart of Oysters; These Eggs must be beaten with some Vinegar, wherein you must put some minced Parsly and Nutmeg finely scraped, and put therein the Oysters in the Pan, which must still be kept stirring least the Liquor make the Eggs curddle, let this all have a good warme on the fire, and serve it up.

To dresse Flounders.

Flea of the black skin, and scowre the Fish over on that side with a Knife, lay them in a dish, and poure on them some Vinegar, and strew good store of Salt, let them lie for halfe an houre; in the mean time set on the fire some water with a little White-Wine, Garlick, and sweet Herbs as you please, putting into it the Vinegar and Salt wherein they lay, when it boyles put in the biggest fish, then the next till all be in; when they are boyled, take them out and drain them very well, then draw some sweet butter thick, and mix with it some Anchoves shred small, which being dissolved in the Butter, poure it on the fish, strewing a little sliced Nutmeg, and minced Oranges and Barberries.

To dresse Snails.

Take Snailes, and put them in a Kettle of water, and let them boyle a little, then take them out, and shake them out of the shels into a Bason; then take some Salt and scoure them very well, and wash them in warme water, untill you find the slime cleane gone from them; then put them into a Cullender and let them draine well, then mince some sweet hearbs, and put them into a Dish with a little Pepper and Sallet-Oyle together, then let them stand an hour or two; then wash the shels very well and dry them, and put into every shell a Snail, and fill up the shell with Sallet-Oyle and herbs, then set them on a gridiron upon a soft fire, and so let them stew a little while, and dish them up warm and serve them up.

To dresse pickle fish.

Wash them well while they are in the shell in salt water, put them into a Kettle over the fire with out water; and stirre them till they are open, then take them out of their shels, and wash them in hot water and salt, then take some of their owne liquor that they have made in the Kettle, a little white wine, butter, vinegar, Spice, Parsley; let all these boyle together, and when it is boyled, take the yolk of three or four Eggs and put into the broth. Scollops may be dressed on this manner or broiled like oysters with Oyle or juyce of Lemons.

To fricate Beefe Pallats.

Take Beefe Pallats after they be boyled very tender, blaunch and pare them clean, season them with fine beaten cloves Nutmeg, Pepper, Salt and some grated bread; then have some butter in a frying Pan, put your pallats therein, and so fricate them till they be browne on both sides, then take them forth and put them on a dish, and put thereto some Gravy of Mutton, wherein two or three Anchoves are dissolved, grate in your sauce a little Nutmeg, wring in the juyce of a Lemon, so serve them.

A Spanish Olio.

Take a peice of Bacon not very fat, but sweet and safe from being rusty, a peice of fresh beefe, a couple of hoggs Eares, and foure feet if they can be had, and if not, some quantity of sheeps feet, (Calves feet are not proper) a joynt of Mutton, the Leg, Rack, or Loyne, a Hen, halfe a dozen pigeons, a bundle of Parsley, Leeks, and Mint, a clove of Garlick when you will, a small quantity of Pepper, Cloves, and Saffron, so mingled that not one of them over-rule, the Pepper and Cloves must be beaten as fine as possible may be, and the Saffron must be first dryed, and then crumble in powder and dissolved apart in two or three spoonfuls of broth, but both the Spices and the Saffron may be kept apart till immediately before they be used, which must not be, till within a quarter of a houre before the Olio be taken off from the fire; a pottle of hard dry pease, when they have first steept in water some dayes, a pint of boyl'd Chesnuts: particular care must be had that the pot wherein the Olio is made, be very sweet; Earthen I thinke is the best, and judgement is to be had carefully both in the size of the Pot, and in the quantity of the Water at the first, that so the Broth may grow afterwards to be neither too much nor too little, nor too grosse, nor too thin; thy meat must be long in boyling, but the fire not too fierce, the Bacon, the Beef, the Pease, the Chesnuts, the Hogs Eares may be put in at the first. I am utterly against those confused Olios into which men put almost all kinds of meats and Roots, and especially against putting of Oyle, for it corrupts the Broath, instead of adding goodnesse to it. To do well, the Broth is rather to be drunk out of a Porringer then to be eaten with a spoon, though you add some smal slices of bread to it, you wil like it the worse. The Sauce for thy meat must be as much fine Sugar beaten smal to powder, with a little Mustard, as can be made to drink the Sugar up, and you wil find it to be excellent, but if you make it not faithfully and justly according to this prescript, but shall neither put Mace, or Rosemary, or Tyme to the Herbs as the manner is of some, it will prove very much the worse.

To make Metheglin.

Take all sorts of Herbs that are good and wholesome, as Balme, Mint, Fennell, Rosemary, Angelica, wilde Tyme, Isop, Burnet, Egrimony, and such other as you think fit; some Field Herbs, but you must not put in too many, but especially Rosemary or any strong Hearb, lesse then halfe a handfull will serve of every sort, you must boyle your Herbs and straine them, and let the Liquor stand till to Morrow and settle them, take off the clearest Liquor, two Gallons and a halfe to one Gallon of Honey, and that proportion as much as you will make, and let it boyle an houre, then set it a cooling as you doe Beere, when it is cold take some very good Ale Barme, and put into the bottome of the Tubb a little and a little as they doe Beere, keeping backe the thicke setling, that lyeth in the bottome of the Vessell that it is cooled in, and when it is all put together, cover it with a Cloth, and let it worke very neere three dayes, and when you mean to put it up, skim off all the Barme clean, put it up into the Vessell, but you must not stop your Vessell very close in three or four dayes, but let it have all the vent, for it will worke, and when it is close stopped, you must looke very often to it, and have a peg in the top to give it vent; when you heare it make a noyse, as it will do, or else it will breake the Vessell; sometime I make a Bag and put in good store of Ginger sliced, some Cloves and Cinnamon, and boyl it in, and other times I put it into the Barrel and never boyle it, it is both good, but Nutmeg and Mace do not well to my Tast.

To make a Sallet of Smelts.

Take halfe a hundred of Smelts, the biggest you can get, draw them and cut off their Heads, put them into a Pipkin with a Pint of White wine, and a Pint of White wine Vinegar, an Onion shred a couple of Lemons, a Race of Ginger, three or foure blades of Mace, a Nutmeg sliced, whole Pepper, a little Salt, cover them, and let them stand twenty foure houres; if you will keep them three or four dayes, let not your Pickle be to strong of the Vinegar, when you will serve them, take them out one by one, scrape and open them as you do Anchoves, but throw away the bones, lay them close one by one, round a Silver dish, you must have the very utmost rind of a Lemon or Orange so small as grated bread and the Parsley, then mix your Lemon Pill, Orange and Parsley together with a little fine beaten Pepper, and strew this upon the dish of Smelts with the meat of a Lemon minced very small, also then power on excellent Sallet Oile, and wring in the juyce of two Lemons, but be sure none of the Lemon-seed be left in the Sallet, so serve it.

To Roast a Fillet of Veal.

Take a Fillet of Beefe which is the tenderest part of the Beast, and lieth only in the inward part of the Surloyne next to the Chine, cut it as big as you can, then broach it on a broach not too big, and be carefull you broach it not thorow the best of the meat, roast it leasurely and baste it with sweet butter. Set a Dish under it to save the Gravy while the Beefe is roasting, prepare the Sauce for it, chop good store of Parsley with a few sweet Herbs shred small, and the yolks of three or foure Eggs, and mince among them the pill of an Orange, and a little Onyon, then boyle this mixture, putting into it sweet butter, Vinegar, and Gravy, a spoonfull of strong broth, when it is well boyled, put it into your beef, and serve it very warm, sometimes a little grosse Pepper or Ginger into your sauce, or a pill of an Orange or Lemon.

To make a Sallet of a cold Hen or a Capon.

Take the breast of a hen or Capon, and slice it as thin as you can in steaks, put therein Vinegar, and a little Sugar as you thinke fit, then take six Anchoves, and a handfull of Capers, a little long, grosse or a carrigon, and mince them together, but not too small, strew them on the Sallet, Garnish it with Lemons, Oranges or barberies, so serve it up with a little salt.

To stew Mushrums.

Take them fresh gathered and cut off the hard end of the stalk, & as you Pil them throw them into a Dish of white Wine, after they have lain half an houre or thereupon draine them from the wine, and put them between two silver Dishes, then set them on a soft fire without any liquor, and when they have so stewed a while, pour away the liquor that comes from them which will be very black, then put your Mushrums into another clean Dish with a sprig or two of Tyme, an Onion whole, four or five cornes of whole Pepper, two or three Cloves, a bit of an Orange, a little Salt, a bit of sweet butter, and some pure gravy of Mutton, cover them, and set them on a gentle fire, so let them stew softly till they be enough and very tender, when you dish them blow off all the fat from them, and take out the Time, spice, and Orange, then wring in the juyce of a Lemon, and grate a little Nutmeg among the Mushrums, tosse them two or three times; put them in a clean dish, and serve them hot to the Table.

The Lord Conway his Lordships receipt for the making of Amber Puddings.

First take the Guts of a young hog, and wash them very clean, and then take two pound of the best hogs fat, and a pound and a halfe of the best Jurden almonds, the which being blancht, take one half of them, & beat them very small, and the other halfe reserve whole unbeaten, then take a pound and a halfe of fine Sugar and four white Loaves, and grate the Loaves over the former composition, and mingle them well together in a bason having so done, put to it halfe an ounce of Ambergreece, the which must be scrapt very small over the said composition, take halfe a quarter of an ounce of levant musk and bruise it in a marble morter, with a quarter of a Pint of orange flower water, then mingle these all very well together, and having so done, fill the said Guts therwith, this Receipt was given his Lordship by an Italian for a great rariety, and has been found so to be by those Ladies of honour to whom his lordship has imparted the said reception.

To make a Partridge Tart.

Take the flesh of four or five Partridges minced very small with the same weight of Beef marrow as you have Partridge flesh, with two ounces of Orangeadoes and green citron minced together as small as your meate, season it with Cloves and Mace and Nutmeg and a little salt and Sugar, mix all together, and bake it in puff past; when it is baked, open it, and put in halfe a Grain of Muske or Amber braid in a Morter or Dish, and with a spoonfull of Rosewater and the juyce of three or four Oranges, when you put all these therein, stir the meat and cover it again, and serve it to the Table.

To keepe Venison all the yeare.

Take the hanch, and parboyle it a while, then season it with two Nutmegs, a spoonfull of Pepper, and a good quantity of salt, mingle them altogether, then put two spoonfulls of white Wine-Vinegar, and having made the Venison full of holes, as you do when you Lard it, when it is Larded, put in the Venison at the holes, the Spice and Vinegar, and season it therewith, then put part into the Pot with the fat side downwards, cover it with two pound of Butter, then close it up close with course Past, when you take it out of the Oven take away the Past, and lay a round Trencher with a weight on the top of it to keep it down till it be cold, then take off the Trencher, and lay the Butter flat upon the Venison, then cover it close with strong white Pepper, if your Pot be narrow at the bottom it is the better, for it must be turned upon a Plate, and stuck with Bayleaves when you please to eat it.

To bake Brawn.

Take two Buttocks and hang them up two or three dayes, then take them down and dip them into hot Water, and pluck off the skin, dry them very well with a clean Cloth, when you have so done, take Lard, cut it in peices as big as your little finger, and season it very well with Pepper, Cloves, Mace, Nutmeg, and Salt, put each of them into an earthen Pot, put in a Pint of Claret wine, a pound of Mutton Suet. So close it with past let the Oven be well heated; and so bake them, you must give them time for the baking according to the bignesse of the Haunches, and the thicknesse of the Pots, they commonly allot seven hours for the baking of them; let them stand three dayes, then take off their Cover, and poure away all the Liquor, then have clarified Butter, and fill up both the Pots, to keep it for the use, it will very well keep two or three moneths.

To roast a Pike.

Take a Pike, scoure off the slime, take out the Entralls, Lard it with the backs of Pickled Herrings, you must have a sharp Bodkin to make the holes, no Larding pins will go thorow, then take some great Oysters, Claret Wine, season it with Pepper, Salt, and Nutmeg, stuff the belly of the Pike with these Oysters, intermix with them Rosemary, Tyme, Winter-Savory, sweet Marjoram, a little Onyon and Garlick, sow these in the belly of the Pike, prepare two sticks about the breadth of a Lath, these two sticks and the Spit must be as broad as the Pike being tyed on the Spit, tye the Pike on, winding Pack-thread about the Pike along, but there must be tyed by the Pack-thred all a long the side of the pike which is not defended by the spit, and the Lathes Rosemary and Bayes, bast the Pike with Butter and Claret-Wine, with some Anchoves dissolved in it, when it is wasted, rip up the belly of the Pike and the Oyster will be the same, but the Herbs which are whole must be taken out.

To sauce Eeles.

Take two or three great Eeles, rubb them in salt, draw out the Guts, wash them very clean, cut them a thwart on both sides found deep, and cut them again cross way, then cut them through in such pieces as you think fit, and put them into a dish with a pint of Wine-Vinegar, and a handfull of Salt, have a kettle over the fire with faire Water, and a bundle of Sweet Herbs, two or thee great Onyons, some Mace, a few Cloves, you must let these lie in Wine-Vinegar and Salt, and put them into boyling liquor, there let them boyl according to Cookery, when enough, take out the Eeles, and drain them from the Liquor, when they are cold, take a pint of Whitewine, boyle it up with Saffron to colour the Wine, then take out some of the Liquor, and put it in an earthen pan take out the onyons and all the herbs, only let the Cloves and Mace remaine, you must beat the Saffron to powder, or else it will not colour.

To make Sausages without skins.

Take a leg of young Pork, two pound of Beef-suet, two handfuls of Sage, two loaves of white bread, Salt and Pepper to your tast, halfe the pork, and halfe the suet, must be very well beat in a stone Morter, the rest cut very small, be sure to cut out all the gresles and Lenets in the pork, when you have mixed these altogether, knead them into a stiffe past with the yolks of two or three Eggs, so rowle them into Sausages.

To dresse a Pike.

Take a Male Pike, rub his skin off whil'st he lives, with bay salt, having well cleared the outside, lay him in a large Dish or Tray, open him so as you break not his gall, cut him according to the size of the fish, in two or three peices, from the head to the taile must be slit, this done, they are to be layd as flat as you can, in a great Dish or Tray, poure upon it halfe a pint of White wine-Vinegar, more or lesse, according to the size of the Fish, then strew upon the inside of the Fish, white Salt plentifully, Bay salt beaten very small is better, whilest this is a doing, let a Skellet with a sufficient quantity of Renish Wine, or good white Wine be pat over the fire, with the Wine, Salt, Ginger, Nutmeg, an Onion, foure or five Cloves of Garlick, a bunch of sweet herbs, *viz.* Sweet Marjoram, Rosemary, peel of halfe a Lemon, let these boyl to the heighth, put in the Pike, with the Vinegar, in such manner as not to quench or allay, if possibly the heat of the Liquor, but the thickest peece first that will aske most boyling, and the Vinegar last of all; while the Pike boyles, take two quarters of a pound of Anchoves, one quarter of very good butter, a Saucer of the Liquor your Pike was boyled in, dissolved Anchoves.

Note that the Liquor, Sauce, the Spice, and the other ingredients must follow the proportion of the Pike; if your Sauce be too strong of the Anchoves, adde more faire water to it. Note also that the Liquor wherein this Pike was dressed, is better to boyle a second Pike therein, then it was at the first.

To dresse Eeles.

Cut two or three Eeles into pieces of a convenient length, set them end-wayes in a pot of Earth, put in a spoolful or two of Water, and to them put some Herbs and Sage chopt small, some Garlick Pepper, and Salt, so let them be baked in an Oven.

To boyle a pudding after the French fashion.

Take a Turkey that is very fat, and being pul'd and drest, Lard him with long pieces of Lard, first wholed in seasoning of Salt, Pepper, Nutmegs, Cloves and Mace, then take one piece of Lard whole in the seasoning, put it into the belly with a sprig of Rosemary and Bayes, sow it very close in a clean cloth, and let it lye all night covered with White-Wine, let it be put into a pot with the same Liquor, and no more, let it be close stopped, then hang it over a very soft and gentle fire, there to continue six houres in a simpering boyle, when it is cold, take it out of the cloth, not before, put it in a Pyeplate, and stick it full of Rosemary and Bayes, so serve it up with Mustard and Sugar, they are wont to lay it on a napkin folded square, and lay it corner wise.

To make a Fricake.

Take three Chickens, and pull off the skins, and cut them into little pieces then put them into water with two or three Onions, and a bunch of Parsly, and when it hath stewed a little, put in some Salt and Pepper, and a pint of white wine, so let them stew till they be enough, then take some Verjuyce, and Nutmegs, and three or foure yolks of Eggs, beat them well together, and when you take off the Chicken, put them into a Frying-Pan altogether with some butter, scald it well over the fire and serve it in.

To make a Dish called Olives.

Take a Fillet of Veale, and the flesh frow the bones, and the fat and skin from either, cut it into very thin slices, beat them with the back of your Knife, lay then abroad on a Dish, season them with Nutmeg, Pepper, Salt and Sugar, chop halfe a pound of Beefe-Suet very small, and strew upon the top of the meat, then take a good handfull of herbs as Parsly, Time, Winter-Savoury, Sorrell, and Spinage, chop them very small, and strew over it, and four Egges with the whites, mingle all these well together with your hands, then roul it up peice by peice, put it upon the spit, roasting it an hour and half, and if it grow dry, baste it with a little sweet Butter, the sauce is Verjuyce or Clarret-Wine with the Gravy of the Meat and Sugar, take a whole Onyon and stew it on a Chafing Dish of coales, and when it tastes of the Onyon, pour the liquor from it on the meat, setting it a while on the coales, and serve it in.

To make an Olive Pye.

This you may take in a Pye, putting Raisins of the Sun stoned and some Currants in every Olive, first strowing upon the meat the whites and yolks of two boyled Eggs shred very small, make your Olives round, and put them into puff paste, when it is halfe baked, put in a good quantity of verjuyce or Clarret wine sweetned with Sugar, putting it in again till it be thorow baked.

The Countesse of RUTLANDS *Receipt of making the rare* Banbury Cake which was so much praised at her Daughters (the right Honourable the Lady Chawerths) wedding.*

Imprimis

Take a peck of fine flower, and halfe an ounce of large Mace, halfe an ounce of Nutmegs, and halfe an ounce of Cinnamon, your Cinnamon and Nutmegs must be sifted through a Searce, two pounds of Butter, halfe a score of Eggs, put out four of the whites of them, something above a pint of good Ale-yeast, beate your Eggs very well and straine them with your yeast, and a little warme water into your flowre, and stirre them together, then put your butter cold in little Lumpes: The water you knead withall must be scalding hot, if you will make it good past, the which having done, lay the past to rise in a warme Cloth a quarter of an hour, or thereupon; Then put in ten pounds of Currans, and a little Muske and Ambergreece dissolved in Rosewater; your Currans must be made very dry, or else they will make your Cake heavy, strew as much Sugar finely beaten amongst the Currans, as you shall think the water hath taken away the sweetnesse from them; Break your past into little pieces, into a kimnell or such like thing, and lay a Layer of past broken into little pieces, and a Layer of Currans, untill your Currans are all put in, mingle the past and the Currans very well, but take heed of breaking the Currans, you must take out a piece of past after it hath risen in a warme cloth before you put in the currans to cover the top, and the bottom, you must roule the cover something thin, and the bottom likewise, and wet it with Rosewater, and close them at the bottom of the side, or the middle which you like best, prick the top and the sides with a small long Pin, when your Cake is ready to go into the Oven, cut it in the midst of the side round about with a knife an inch deep, if your Cake be of a peck of Meale, it must stand two hours in the Oven, your Oven must be as hot as for Manchet.

An excellent Sillabub.

Fill your Sillabub-pot with Syder (for that is the best for a Sillabub) and good store of Sugar and a little Nutmeg; stir it well together, put in as much thick Cream by two or three spoonfuls at a time, as hard as you can, as though you milke it in, then stir it together exceeding softly once about, and let it stand two hours at least ere it is eaten, for the standing makes the Curd.

To Sauce a Pig.

Take a faire large Pigge and cut off his Head, then slit him through the midst, then take forth his bones, then lay him in warme water one night, then Collar him up like Brawne; then boyle him tender in faire water, and when he is boyled put him in an earthen Pot or Pan into Water and Salt, for that will make him white, and season the flesh, for you must not put Salt in the boyling, for that will make it black, then take a quart of the same broth, and a quart of white wine; boyl them together to make some drink for it, put into it two or three Bay leaves, when it is cold uncloathe the Pig, and put it into the same drink, & it will continue a quarter of a year. It is a necessary Dish in any Gentlemans House; when you serve it in, serve it with greene Fennell, as you doe Sturgion with Vinegar in Saucers.

To make a Virginia Trout.

Take Pickled Herrings, cut off their Heads, and lay the bodies two dayes and nights in water, then wash them well, then season them with Mace, Cinamon, Cloves, Pepper, and a little Red Saunders, then lay them close in a pot with a little onyon strewed small upon them, and cast between every Layer; when you have thus done, put in a pint of Clarret-Wine to them, and cover them with a double paper tyed on the pot, and set them in the oven with houshould-bread. They are to be eaten cold.

To make a fat Lamb of a Pig.

Take a fat Pig and scald him, and cut off his head, slit him and trusse him up like a Lamb, then being slit through the middle, and flawed, then parboyle him a little, then draw him with parsley as you do a Lamb, then roast it and dridge it, and serve it up with Butter, Pepper, and Sugar.

To make Rice Pancakes.

Take a pound of Rice, and boyle it in three quarts of water till it be very tender, then put it into a pot covered close, and that will make a Jelly, then take a quart of Cream or new Milk, put it scalding hot to the Rice, then take twenty Eggs, three quarters of a pound of melted Butter, a little Salt, stirre all these well together, put as much flowre to them as will make them hold frying, they must be fryed with Butter, they must be made overnight, best.

Mrs. Dukes Cake.

Take a quarter of a peck of the finest flour, a pint of Cream, ten yolks of Eggs well beaten, three quarters of a pound of butter gently melted, pour on the floure a little Ale-yeast, a quarter of a pint of Rose water, with some Muske, and Amber-grece dissolved in it, season all with a penny worth of Mace and Cloves, a little Nutmeg finely beaten, Currans one pound and a halfe, Raisins of the Sun stoned, and shred small one pound, Almonds blanch'd and beaten, halfe a pound, beat them with Rosewater to keep them from Oyling: Sugar beaten very small, half a pound; first mingle them, knead all these together, then let them lie a full houre in the Dough together, then the Oven being made ready, make up your Cake, let not the oven be too hot, nor shut up the mouth of it too close, but stir the Cake now and then that it may bake all a like, let it not stand a full hour in the Oven. Against you draw it have some Rose water and Sugar finely beaten, and well mixed together to wash the upper side of it, then set it in the Oven to dry, when you draw it out, it will shew like Ice.

To make fine Pancakes fryed without Butter, or Lard.

Take a Pint of Creame, six new layd Eggs, beat them very well, put in a quarter of a Pound of Sugar, one Nutmeg or beaten mace which you please, as much floure as will thicken them almost as thick as for ordinary Pancakes, your Pan must be cleane wiped with a Cloth, when it is reasonably hot, put in your Butter, or thick or thin as you please, to fry them.

To pot Venison.

Take a haunch of Venison not hunted, and bone it, then take three ounces of Pepper beaten, twelve Nutmegs, with a handfull of Salt, and mince them together with Wine Vinegar, then wet your Venison with Wine Vinegar and season it, then with a knife make holes on the lean sides of the Hanch, and stuff it as you would stuff Beef with Parsley, then put it into the Pot with the fat side downward then clarifie three pound of Butter, and put it thereon, and Past upon the Pot, and let it stand in the Oven five or six hours, then take it out, and with a vent presse it down to the bottom of the Pot, and let it stand till it be cold, then take the Gravy of the top of the Pot and melt it, and boyle it halfe away and more, then put it in again with the Butter on the top of the Pot.

To make a Marchpan; to Ice him, &c.

Take two pound of Almonds blanched, & beaten in a stone Morter till they begin to come to a fine Past, and take a pound of sifted Sugar, and put it in the Morter with the Almonds, and so leave it till it come to a perfect Past, putting in now and then a Spoonfull of Rosewater to keep them from Oyling; when you have beaten them to a perfect Past cover the Marchpan in a sheet, as big as a Charger, and set an edge about as you do about a Tart, and a bottome of wafers under him; thus bake it in an oven or baking pan, when you see your marchpan is hard and dry, take it out and Ice him with Rosewater and sugar being made as thick as butter for Fritters; so spread it on him with a wing-feather; so put it into the Oven againe, and when you see it rise high, then take it out and garnish it with some pretty conceits made part of the same stuff, stick long cumfets uprigh in him so serve it.

To make Jelly the best manner.

Take a Leg of Veale, and pare away the fat as clean as you can, wash it throughly, let it lie soaking a quarter of an hour or more, provided you first breake the bones, then take foure Calves feet, scald off the hair in boyling water, then slit them in two and put them to your Veale, let them boyle over the fire in a brasse pot with two Gallons of water or more acording to the proportion of your Veale, scum it very clean and often; so let it boyle till it comes to three Pintes or little more, then strain it through a cleane strainer, into a Bason, and so let it stand till it be through cold and well jellied, then cut it in peices with a Knife, and pare the top and the bottome of them, put it into a Skellet, take two ounces of Cynamon broken very small with your hand, three Nutmegs sliced, one race of Ginger, a large Mace or two, a little quantity of Salt, one Spoonfull of Wine Vinegar, or Rose-Vinegar, one pound and three quarters of Sugar, a Pint of Renish-wine, or white Wine, and the Whites of fifteen Eggs, well beaten; put all these to the Jelly, then set it on the fire, and let it seeth two or three walmes, ever stirring it as it seeths, then take a very clean Jelly bag, wash the bottom of it in a little Rose water, and wring it so hard that their remaine none behind, put a branch of Rosemary in the bottom of the bag, hang it up before the fire over a Bason; and pour the Jelly-bag into the Bason, provided in any case you stir not the Bag, then take Jelly in the Bason and put it into your bag again, let it run the second time, and it will be very much the clearer; so you may put it into your Gallypots or Glasles which you please, and set them a cooling on bay salt, and when it is cold and stiffe you may use it at your pleasure, if you will have the jelly of a red colour use it as before, onely instead of Renish wine, use Claret.

To make poore knights.

Cut two penny loaves in round slices, dip them in half a pint of Cream or faire water, then lay them abroad in a dish, and beat three Eggs and grated Nutmegs and sugar, beat them with the Cream then melt some butter in a frying pan, and wet the sides of the toasts and lay them in on the wet side, then pour in the rest upon them, and so fry them, serve them in with Rosewater, sugar and butter.

To make Shrewsbury Cakes.

Take two pound of floure dryed in the Oven and weighed after it is dryed, then put to it one pound of Butter that must be layd an hour or two in Rose-water, so done poure the Water from the Butter, and put the Butter to the flowre with the yolks and whites of five Eggs, two races of Ginger, and three quarters of a pound of Sugar, a little salt, grate your spice, and it well be the better, knead all these together till you may rowle the past, then roule it forth with the top of a bowle, then prick them with a pin made of wood, or if you have a comb that hath not been used, that will do them quickly, and is best to that purpose, so bake them upon Pye plates, but not too much in the Oven, for the heat of the Plates will dry them very much, after they come forth of the Oven, you may cut them without the bowles of what bignesse or what fashion you please.

To make beef like red Deer to be eaten cold.

Take a buttock of beef, cut it the long wayes with the grain, beat it well with a rowling pin, then broyl it upon the coals, a little after it is cold, draw it throw with Lard, then lay in some white wine Vinegar, Pepper, Salt, Cloves, Mace and Bay-leaves, then let it lie three or four dayes, then bake it in Rye past, and when it is cold fill it up with butter, after a fortnight it will be eaten.

To make puffs.

Take a pint of Cheese Curds and drain them dry, bruise them small with the hand, put in two handfulls of floure, a little Sugar, three or four yolks of Egs, a little Nutmeg and Salt, mingle these together and make them little, like eyes, fry them in fresh butter, serve them up with fresh Butter and Sugar.

To make a hash of Chickens.

Take six Chickens, quarter them, cover them almost with water, and season them with Pepper and Salt, and a good handfull of minced Parsly, and a little white-wine, when they are boyled enough, put six Eggs onely the yolks, put to them a little Nutmeg and Vinegar, give them a little wame or two with the Chickens, pour them altogether into the Dish and serve them in, when you put on the Eggs, and a good piece of Butter.

To make an Almond Caudle.

Take three pints of Ale, boyle it with Cloves, Mace and sliced Bread into it, then have ready beaten a pound of blanched Almonds stamped in a Mortar with a little white-wine, then strain them out with a pint of white-wine, thick your Ale with it, sweeten it as you please, and be sure you skim the Ale well when it boyles.

To make scalding Cheese towards the latter end of May.

Take your Evening Milke and put it into Boules, or Earthen Pans, then in the Morning, fleet off the Cream in a Boule by it selfe, put the fleet Milke into a Tub with the Morning Milk, then put in the nights Cream, and stir it together, and heat the Milk, and put in the Rennet; as for ordinary new Milk Cheese, it is to be made thick; when the Cheese is come, gather the Curd into a Cheese-cloath, and set the Whey on the fire till it be seething hot, put the Cheese in a Cloth into a Killar that hath a wafle in the bottome, and poure in the hot Whey, then let out that, and put in more till your Curd feele hard, then break the Curd with your hands, as small as you can, and put an handfull of Salt to it then put it into the Fat, thrune it at noon and at night, and next day put it into a Trough where Cheese is salted every day, and turne it as long as any will enter, then lay it on a Table or Shelfe all Summer; if you will have it mellow to eate within an yeare, it must be laid in Hay in the Spring; if to keep two yeares, let it dry on a Shelfe out of the Wind all the next Summer, and in Winter lay them in Hay a while, or lay them close one to another; I seldome lay any in Hay, I turne and rub them with a rotten cloth especially when they are old, once a week least they rot.

To Pickle Purslaine.

Take Purslaine, stalks and all, boyl them tender in faire Water, then lay them drying upon linning Cloaths, then being dryed, put them into the Galley-pots and cover them with wine Vinegar mixt with Salt, and not make the Pickle so strong as for Cucumbers.

FINIS.

THE TABLE TO the Compleat COOK.

To make a Posset the Earle of Arundels *way.*
To boyle a Capon larded with Lemons.
To bake Red Deer.
To make fine Pancakes fryed without Butter or Lard.
To dresse a Pig the French manner.
To make a Steak Pye with a French Pudding in the Pye.
An excellent way for dressing Fish.
To Fricate Sheeps feet.
To Fricate Calves Chaldrons.
To Fricate Campigneons.
To make buttered Loaves.
To marine Carps, Mullet, Gormet, Rochet, or Wale.
To make a Calves Chaldron Pye.
To make a Pudding of Calves Chaldron.
To make a Banbury *Cake.*
To make a Devonshire *White Pot.*
To make Rice cream.
To make a very good Oxfordshire *cake.*
To make a Pompion Pye.
To make the best Sausages.
To boyle fresh fish.
To make friters.
To make loaves of Cheese curd.

To make fine Pyes after the French fashion.
A singular good receipt for making a Cake.
To make a great curd Loafe.
To make buttered Loaves of Cheese curds.
To make Cheese Loaves.
To make Puffe.
To make Elder Vinegar.
To make good Vinegar.
To make a collar of Beefe.
To make an Almond Pudding.
To boyle Creame with French Barly.
To make Cheese cakes.
To make a quaking Pudding.
To pickle Cucumbers.
To pickle broom buds.
To keep Quinces all the yeare.
To make a goosberry fool.
To make an Oatmeale pudding.
To make a green Pudding.
To make good Sausages.
To make toasts.
A Spanish cream.
To make clouted cream.
A good cream. To make Pyramids cream.
To make a sack cream.
To boyl Pigeons.
To make an apple tansey.
A french barly cream.

To make a Chicken or Pigeon Pye.

To boyle a capon or hen.

To make bals of Veal.

To make Mrs. Shelleyes cake.

To make Almond Jumbals.

To make cracknels.

To pickle Oysters.

To boyl cream with codlings.

To make the lady Abergaveers Cheese.

To dresse snails.

To boyl a rump of Beefe after the French fashion.

An excellent way of dressing fish.

To make fritters of Sheeps feet.

To make dry Salmon calvert in the boyling.

To make bisket bread.

To make an Almond pudding.

To make an Almond caudle.

To make Almond bread.

To make Almond cakes.

Master Rudstones posset.

To boyle a capon with Ranioles.

To make a bisque of carps.

To boyle a Pike and an Eele together.

To make an outlandish dish.

To make a Portugal dish.

To dresse a dish of Hartichockes.

To dresse a Fillet of Veal the Italian way.

To dresse soals.

To make furmity.

To make a patis or cabbage cream.

To make Pap.

To make Spanish Pap.

To poach Eggs.

A pottage of beefe Pallats.

The Jacobins pottage

To salt a Goose.

A way of stewing Chickens or Rabbets.

A pottage of Capons.

A Carp pye.

To boyle Ducks after the French fashion.

To boyle a goose with sausages.

To fry Chickens.

To make a battalia Pye.

To make a Chicken pye.

To make a pye of a Calves head.

To make Cream with Snow.

To make minced Pyes.

To drye Neates tongues.

To make jelly of harts horn.

To make Chickens fat in four or five dayes.

To make Angelot.

A Persian dish.

To roast a shoulder of Mutton.

To roast a leg of Mutton to be eaten cold.

To roast Oysters.

To make a Sack Posset.

Another

To make a Sack Posset without Milk or Creame.

To make a stump pye.

To make Mrs. **Leed** *Cheese Cakes.*

To make taffaty tarts

To make fresh Cheese

To make Sugar Cakes or Jumballs

To hash a shoulder of Mutton

To dresse Flounders or Plaice with Garlick and Mustard

A turkish dish

To dresse a Pike

To dresse Oysters

To dresse Flounders

To dresse Snailes

To dresse pickle fish

To fricate beef Pallats

A Spanish Olio

To make a Spanish Olio.

To make Metheglin

To make a sallet of smelts

To roast a Fillet Beefe

To make a sallet of a cold Hen or Capon.

To stew Mushrumps

The Lord **Conway** *his receipt for the makeing of Amber-puddings*

To make a Partridge tart

To keep venison all the yeare

To make Brawn

To roast a Pike

To sauce Eeles

To make sausages without skins

To dresse a Pike.

To dresse Eeles

To boyle a pudding after the French fashion,

To make a fricate

To make a dish called Olives

To make an Olive Pye

The Countesse of Rutlands *Receipt of makeing a rare* Banbury Cake

An excellent Syllabub

To sauce a Pig

To make a Virginia trout

To make a fat Lamb of a Pig.

To make Rice pancakes

Mrs. Dukes *Cakes.*

To make fine Pancakes.

To pot Venison

To make a Marchpan to ice him

To make jelly the best manner

To make poor Knights

To make Shrewsberry Cakes

To make Beefe like Red Deere to be eaten Cold

To make Puffe

To make a hash of Chicken

To make an Almond Caudle

To make scalding Cheese towards the latter end of May

To pickle purslain

FINIS.

Courteous READER, *these Books following are Printed for* Nath. Brook, *and are to be sold at his Shop at the Angell in* Cornhill.

* * * * *

Excellent Tracts in Divinity, Controversies, Sermons, Devotions.

The Catholique History collected and gathered out of Scripture, Councels, and Antient Fathers, in answer to Dr. *Vanes* Lost Sheep returned home: by *Edward Chesensale* Esq; *Octavo.*

2. Bishop *Morton* on the Sacrament, in *Folio.*

3. The Grand Sacriledge of the Church of *Rome*, in taking away the sacred Cup from the Laity at the Lords Table; by Dr. *Featly* D.D. *Quarto.*

4. The Quakers Cause at second hearing, being a full answer to their Tenets.

5. Re-assertion of Grace: *Vindiciae Evangelii,* or the Vindication of the Gospell: a reply to Mr. *Anthony Burghess Vindiciae Legis,* and to Mr. *Ruthford*: by *Robert Town.*

6. Anabptists anatomized and silenced: or a dispute with Master *Tombs*, by Mr. *J. Crag*: where all may receive cleare satisfaction in that controversie, the best extant. *Octavo.*

7. A Glimpse of Divine Light, being an explication of some passages exhibited to the Commissioners of *White Hall* for Approbation of Publique Preachers, against *John Harrison* of *Land Chap. Lancash.*

8. The Zealous Magistrate: a Sermon by *T. Threscos. Quarto.*

9. New Jerusalam, in a Sermon for the society of Astrologers, *Quarto.* in the year 1651.

10. Divinity no enemy to Astrology: A Sermon for the society of Astrologers, in the year 1653. by D. *Thomas Swadling.*

11. *Britannia Rediviva,* a Sermon before the Judges, *August* 1648. by *J Shaw* Minister of *Hull.*

12. The Princess Royal, in a Sermon before the Judges, *March* 24. by *J Shaw*.

13. Judgement set, and books opened, Religion tried whether it be of God or Man, in severall Sermons: by *J Webster, Quarto*.

14. Israels Redemption, or the Prophetical History of our Saviours Kingdome on Earth: by *K. Marton*.

15. The Cause and Cure of Ignorance, Error and Prophaness: or a more hopefull way to Grace and Salvation: by K. *Young, Octavo*.

16. A Bridle for the Times, tending to still the murmuring, to settle the wavering, to stay the wandring, and to strengthen the fainting: by *J Brinsley* of *Yarmouth*.

17. Comforts against the fear of death; wherein are discovered severall Evidences of the work of Grace: by *J Collins* of *Norwich*.

18. *Jacobs* Seed: or, the excellency of seeking God by prayer, by *Jer Burroughs*.

19. The form of Practical Divinity; or, the grounds of Religion in a Chatechistical way, by Mr. *Christopher Love* late minister of the gospel: a useful piece.

20. Heaven and Earth shaken; a Treatice shewing how Kings and Princes, their Governments are turned and changed, by *J Davis* Minister in *Dover*: admirably useful and seriously to be considered in these times.

21. The Treasure of the Soul; wherein we are taught, by dying to sin, to attain to the perfect love of God.

22. A Treatise of Contestation fit for these sad & troublesome times by *J. Hall* Bishop of *Norwich*.

23. Select thoughts: or, choice helps for a pious spirit, beholding the excellency of her Lord Jesus; by *J. Hall* Bishop of *Norwich*.

24. The Holy Order, or Fraternity of Mourners in Zion; to which is added, Songs in the night, or chearfulness under afflictions; by *J. Hall* Bishop of *Norwich*.

25. The Celestial Lamp, enlightening every distressed Soul from the depth of everlasting darkness; by *T. Fetisplace*.

Admirable and learned Treatises of Occult Sciences in Philosophy, Magick, Astrology, Geomancy, Chymistry, Physiognomy, and Chyromancy.

26. Magick & Astrology vindicated by H. *Warren*

27. *Lux Veritatis*, Judicall Astrology vindicated and demonology confuted; by *W. Ramsey* Gent.

28. An Introduction to the Tentonick Philosophy; being a determination of the Original of the Soul: by *C. Hotham* Fellow of *Peter-House* in *Cambridge*.

29. *Curnelius Agrippa*, his fourth book of Occult Philosophy, or Geomancy: Magical Elements o *Peter de Abona*, the nature of Spirits: made English by *R Turner*.

30. *Paracelsus* Occult Philosophy, of the Misteries of Nature, and his Secret Alchimy.

31. An Astrological Discourse with Mathematical Demonstrations; proving the influence of the Planets and fixed Stars upon Elementary Bodies: by Sir *Chr. Heydon* Knight.

32. *Merlinus Anglicus Junior*; the English Merlin revived: or a Prediction upon the Affairs of Christendome, for the year 1644, by *W. Lilly*.

33. Englands Prophetical Merlin; foretelling to all Nations of *Europe*, till *1663*. the actions depending upon the influences of the Conjunction of *Saturn* and *Jupiter* 1642. by *W. Lilly*.

34. The Starry messenger: or an Interpretation of that strange apparition of three Suns seen in *London*, the 19 of *November* 1644, being the birthday of King *Charles*: by *W. Lilly*.

35. The Worlds Catastrophe: or *Europes* many Mutations, untill 1666, by *W. Lilly*.

36. An Astrological Prediction of the Occurrences in *England*; part of the years 1648, 1649, 1650. by *W. Lilly*.

37. Monarchy or no Monarchy in *England*: the Prophesie of the white King, *Grebner* his Prophesie, concerning *Charles*, Son of *Charles*, his greatness; illustrated with several Hieroglyphicks: by *W. Lilly*.

38. *Annus Tenebrosus*, or the Dark Year, or Astrological Judgements upon two Lunary Eclipses, and one admirable Eclipse of the Sun in *England* 1652. by *W. Lilly*.

39. An easie and familiar Method, whereby to judge the effects depending on Eclipses: by *W. Lilly*.

40. Supernatural Sights and Apparitions seen in *London, June 30* 1644. by *W. Lilly*: as also all his Works in a volumn.

41. *Catastrophe Magnatum*: an Ephemerides for the year 1652. by *N. Culpeper*.

42. *Teratologia*; or, a discovery of Gods Wonders, manifested by bloody raine and waters, by *I.S.*

43. Chyromancy; or the Art of divining by the lines egraven in the hand of man, by dame nature in 19. Genitures; with a Learned Discourse of the Soul of the World; by *G. Wharton* Esq.

44. The admired piece of Physiognomy, and Chyromancy, Metoposcopy, and Simmetricall Proportions, and Signal moles of the Body, and Interpretation of Dreams: to which is added the art of Memory, illustrated with figures: by *R. Sanders*, in *Folio*.

45. The no less exquisite then admirable Work, *The atrum chemicum Britannicum*; containing several Poetical pieces of our famous English Philosophers, who have written the Hermitique Mysteries in their own antient Language; faithfully collected into one Volumn, with Annotations thereon: by the Indefatigable Industry of *Elias Ashmole* Esq; illustrated with Figures.

Excellent Treatises in the Mathematicks, Geometry of Arithmetick, Surveying, and other Arts or Mechannicks.

46. The incomparable Treatise of *Tactometria, seu. Tetagmenometria*; or, the Geometry of Regulars, practically proposed, after a new and most expeditious manner, (together with the Natural or Vulgar, by way of Mensural comparison) and in the Solids, not only in respect of Magnitude or Demension, but also of Gravity or Ponderosity, according to any Metall assigned: together with useful experiments

of Measures & Weights, observations on Gauging, useful for those are practised in the Art Metricald: by *T. Wibard*.

47. *Tectonicon*, shewing the exact measuring of all manner of Land, Squares, Timber Stone, Steeples, Pillars, Globes; as also the making and use of the Carpenters Rule &c. fit to be known by all Surveyors, Land-meters, Joyners, Carpenters, and Masons: by *L. Digges*.

48. The unparalleld work for ease & expedition, instituted, The exact Surveyor: or, the whole Art of Surveying of Land, shewing how to plot all manner of Grounds, whether small Inclosures, Champain, Plain, Wood-Lands, or Mountains, by the Plain Table; as also how to finde the Area, or Content of any Land, to Protect, Reduce or Divide the same; as also to take the Plot or Cart, to make a map of any Manner, whether according to *Rathburne*, or any other Eminent Surveyors Method: a Booke excellently useful for those that sell, purchase, or are otherwise employed about Buildings; by *J. Eyre*.

49. *Moor's* Arithmetick: discovering the secrets of that Art, in Number and Species; in two Books, the first teaching by precept and example, the Operations in numbers, whole and broken. The Rules of Practice, Interest, and performed in the more facil manner by Decimals, then hitherto hath been published; the excellency and new practice and use of Logarithmes, *Nepayres Bones*. The second the great Rule of *Algebra*, in Species, resolving all Arithmetical Questions by Supposition.

50. The golden Treatise of Arithmetick, Natural and Artificial, or Decimals; the Theory & Practice united in a simpathetical Proportion, betwixt lines and Numbers, in their Quantities and Qualities, as in respect of Form, Figure, Magnitude and Affection; demonstrated by Geometry, illustrated by Calculations, and confirmed with variety of Examples in every Species; made compendious and easie for Merchants, Citizens, Sea-men, Accomptants, &c. by *Th. Wilsford* Corrector of the last Edition of Record.

51. Semigraphy, or the Art of Short-Writing, as it hath been proved by many hundreds in the City of *London*, and other places, by them practised, and acknowledged to be the easiest, exactest, and swiftest method; the meanest capacity by the help of this Book,

with a few hours practice, may attaine to a perfection in this Art: by *Jer. Rich* Author and Teacher thereof, dwelling in *Swithings Lane* in London.

52. Milk for Children; a plain and easie method teaching to read and write, usefull for Schools and Families, by *L. Thomas*, D.D.

53. The Painting of the Ancients; the History of the beginning, progress, and consummating of the practice of that noble Art of Painting; by *F. Junius*

Excellent and approved Treatises in Physick, Chyrurgery, & other more familiar Experiments in Cookery, Preserving, &c.

54. *Culpeper's Semiatica uranica,* his Astrological judgement of Diseases from the decumbiture of the sick, much enlarged: the way and manner of finding out the cause, change, and end of the Disease; also whether the sick be likely to live or dye, & the time when recovery or death is to be expected, according to the judgement of *Hipocrates,* and *Hermes Trismegistus;* to which is added Mr. *Culpeper's* censure of Urines.

55. *Culpeper's* last Legacy, left to his Wife for the publick good, being the choicest and most profitable of those secrets in Physick and Chyrurgery; which whilst he lived, were lockt up in his breast, and resolved never to be published till after his death.

56. The Yorkshire Spaw; or the virtue and use of that water in curing of desperate diseases, with directions and rules necessary to be considered by all that repair thither.

57. Most approved Medicines and Remedies for the diseeses in the body of Man: by *A. Read* Dr. in Physick.

58. The Art of Simpling: an introduction to the knowledg of gathering of Plants, wherein, the definitions, divisions, places, descriptions, differences, names, virtues, times of gathering, uses, tempratures of them are compendiously discoursed of: also a discovery of the lesser World, by *W. Coles.*

59. *Adam* in Eden, or Natures Paradise: the History of Plants, Herbs and Flowers, with their several original names, the places

where they grow, their descriptions and kindes, their times of flourishing and decreasing; as also their several signatures, anatomical appropriations, and particular physical virtues; with necessary Observations on the seasons of planting and gathering of our English plants. A work admirably useful for Apothecaries, Chyrurgeons, and other Ingenuous persons, who may in this Herbal finde comprised all the English physical simples, that *Gerard* or *Parkinson*, in their two voluminous Herbals have discoursed of, even so as to be on emergent occasions their own physitians, the ingredients being to be be had in their own fields & gardens, Published for the general good by *W. Coles* M.D.

60. The Compleat Midwive's practice, in the high & weighty concernments of the body of Mankinde: or perfect Rules derived from the experiences and writings, not onely of our English, but the most accomplisht and absolute practices of the French, Spanish, Italians, and other Nations; so fitted for the weakest capacities, that they may in a short time attain to the knowledge of the whole art; by *Dr.* T.C. with the advice of others, illustrated with Copper figures.

61. The Queens Closet opened: incomparable secrets in Physick, Chyrurgery, Preserving, Candying, and Cookery; as they were presented to the queen by the most experienced persons of our times; many whereof were honour'd with her own practice.

Elegant Treatises in Humanity, History, Romances, & Poetry.

62, Times Treasury, or Academy, for the accomplishment of the English Gentry in Arguments of Discourse, Habit, Fashion, Behaviour, &c. all summed up in Characters of Honour: by *R. Brathwait*, Esq.

63. *Oedipus*, or the Resolver of the secrets of love, and other natural Problemes, by way of Question and Answer.

64. The admirable and most impartial history of *New England*, of the first plantation there, in the year 1628. brought down to these times; all the material passages performed there, exactly related.

65. The Tears of the Indians: the History of the bloody and most cruel proceedings of the Spaniards in the Islands of *Hispaniola, Cuba,*

Jamaica, Mexico, Peru, and other places of the West Indies; in which to the life, are discovered the tyrannies of the Spaniards, as also the justnesse of our War so successfully managed against them.

66. The illustrious Sheperdess. The Imperious Brother: written originally in Spanish by that incomparable wit, *Don John Perez de Montalban;* translated at the request of the Marchioness of *Dorchester,* and the countess of *Strafford:* by E.P.

67. The History of the Golden Ass, as also the Loves of Cupid and his Mistress *Psiche:* by *L. Apulcius* translated into English.

68. The unfortunate Mother: a tragedy by *T.N.*

69. The Rebellion, a Comedy by *T. Rawlins.*

70. The tragedy of *Messalina* the insatiate Roman Empress: by *N. Richards.*

71. The Floating Island: a Trage-Comedy, acted before the King, by the students of Christs-Church in *Oxon;* by that renowned wit, *W. Strode* the Songs were set by Mr. *Henry Lawes.*

72. *Harvey's* Divine Poems: the History of *Balaam,* of *Jonah,* and of St. *John* the Evangelist.

73. *Fons Lachrymarum,* or a Fountain of Tears; the lamentations of the Prophet *Jeremiah* in verse, with an Elegy on Sir *Charles Lucas;* by *I. Quarles.*

74. Nocturnal Lucubrations, with other witty Epigrams and Epitaphs; by *R. Chamberlain.*

75. The admirable ingenuous Satyr against Hypocrites.

Poetical, with several other accurately ingenious Treatises, lately Printed.

76. Wits Interpreter, the English Parnassus: or a sure Guide to those admirable accomplishments that compleat the English Gentry, in the most acceptable qualifications of Discourse or Writing. An Art of Logick, accurate Complements, Fancies, and Experiments, Poems, Poetical Fictions, and *All-a-Mode* Letters by J.C.

77. Wit and Drollery; with other Jovial Poems: by sir *I.M.M.L.M.S.W.D.*

78. Sportive wit, the Muses Merriment; a New Sprint of Drollery; Jovial Fancies, &c.

79. The Conveyancer of Light, or the Compleat Clerk, & Scriviners Guide; being an exact draught of all Presidents and Assurances now in use; as they were penned, and perfected by diverse learned Judges, eminent Lawyers, & great Conveyancers, both ancient and modern: whereunto is added a Concordance from *K. Rich 3*. to this present.

80. *Themis Aurea*, The Daws of the Fraternity of the Rosie Cross; in which the occult secrets of their Philosophical Notions are brought to light: written by *Count Mayerus*, and now Englisht by *T.H.*

82. The Iron Rod put into the Lord Protectors hand; a phrophetical Treatise.

83. *Medicina magica tamen Physica*; Magical but Natural Physick: containing the general cures of infirmities and diseases belonging to the bodies of men, as also to other animals and domistick creatures, by way of Transplantation: with a description of the most excellent Cordial out of Gold; by *Sam. Boulton of Salop*.

84. *I. Tradiscan's* Rareties, published by himself.

85. The proceedings of the high Court of Justice against the late King Charles, with his Speech upon the Scaffold, and other proceedings, *Jan.* 30, 1648.

86. The perfect Cook; a right Method in the Art of Cookery, whether for Pastery, or all other manner af *All-a-mode* Kick shaws; with the most refined ways of dressing of Flesh, Fowl, or Fish; making of the most poinant Sawces, whether after the French or English manner, together with fifty five ways of dressing of Eggs; by *M. M.*

Admirable usefull Treatises Newly Printed.

87. The Expert Doctors Dispensatory: the whole Art of Phisick restored to Practise: the Apothecaries Shop, and Chyrurgeons Closet opened; with a Survey, as also a Correction of most Dispensatories

now extant, with a Judicious Cencure of their defects; & a supply of what they are deficient in: together with a learned account of the virtues and quantities, and uses of Simples, and Compounds; with the Symptoms of Diseases; as also prescriptions for their several cures: by that renowned *P. Morellus* Physician to the King of *France*; a work for the order, usefulness, and plainness of the Method, not to be parallel'd by any Dispensatory, in what Language soever.

88. Cabinet of Jewels, Mans Misery, Gods Mercy, Christs Treasury, &c. In eight Sermons; with an Appendix of the nature of Tithes under the Gospel; with an expediency of Marriage in Publique Assemblies, by *I. Crag* Minister of the Gospel.

89. Natures Secrets; or the admirable and wonderful History of the generation of Meteors; discribing the Temperatures of the Elements, the heights, magnitudes, and influences of Stars; the causes of Comets, Earthquakes, Deluges, Epidemical Diseases and Prodigies of precedent times, with presages of the weather, and Descriptions of the Weather-glass: by *T. Wilsford*.

90. The Mysteries of Love and Eloquence; or the Arts of Wooing and Complementing; as they are managed in the *Spring Garden, Hide-Park, the New Exchange*, and other Eminent Places. A work in which are drawn to the Life and Deportments of the most Accomplisht Persons; the Mode of their Courtly Entertainments, Treatment of their Ladies at Balls, their accustomed Sports, Drolls & Fancies; the witchcrafts of their perswasive language, in their approaches, or other more secret dispatches, *&c.* by *E.P.*

91. *Helmont* disguised; or the vulgar errors of imperical and unskilful practicers of Physick confuted; more especially as they concern the cures of Feavers, the Stone, the Plague, and some other Diseases by way of Dialogue; in which the chief rarities of Physick are admirably discoursed by *I.T.*

Books in the Press, and ready for Printing.

1. The Scales of Commerce and Trade: by *T. Wilsford*.

2. Geometry demonstrated by Lines & Numbers; from thence, Astronomy, Cosmgraphy, and Navigation proved and delineated by the Doctrine of Plane and Spherical Trangles: by *T. Wilsford*.

3. The English Annals, from the Invasion made by Julius Cesar to these times: by *T. Wilsford*.

4. The Fool tranformed: a Comedy.

5. The History of *Lewis* the Eleventh King of *France*: a Trage-Comedy.

6. The chast woman against her will: a Comedy.

7. The Tooth-Drawer: a Comedy.

8. Honour in the end: a Comedy.

9. The Tell Tale: a Comedy

10. The History of *Donquixiot*, or the Knight of the illfavour'd Face: a Comedy.

11. The fair Spanish Captive: a Trage-Comedy.

12. Sir *Kenelm Digby* & other persons of Honour, their rare and incomparable secrets of Physick, Chyrurgery, Cookery, Preserving, Conserving, Candying, distilling of Waters, extraction of Oyls, compounding of the costliest Perfumes, with other admirable Inventions, and select Experiments, as they offered themselves to their Observations, whether here or in Forrein Countreys.

13. The so much desired & deeply learned Commentary on *Psalme* 15. by that reverend and eminent Divine Mr. *Christopher Carthwright* Minister of the Gospel in *York*.

14. The Soul's Cordial in two treatises, the first teaching how to be eased of the guilt of sin, the second, discovering advantages by Christs ascension: by that faithful labourer in the Lord's vineyard Mr. *Christopher Love*, late Parson of *Laurance Jury*: the third volumn.

15. Jacobs seed, the excellency of seeking God by prayer, by the late reverend divine *I. Burroughs*.

16. The Saints Tombe-Stone: or the Remains of the Blessed: A plain Narrative of some remarkable passages, in the Holy Life, & Happy Death, of Mrs. *Dorothy Shaw*, wife of Mr. *John Shaw* Preacher

of the Gospel at *Kingston* on *Hull* collected by her dearest friends especially for her sorrowful Husband and six Daughters consolation and invitation.

17. The Accomplisht Cook, the mistery of the whole art of Cookery, revealed in a more easie and perfect method then hath been publisht in any language: Expert and ready wayes for the dressing of flesh, fowl and fish, the raising of pastes, the best directions for all manner of Kickshaws and the most poinant Sauces, with the termes of Carveing and Sewing: the Bills of fare, an exact account of all dishes for the season, with other All-a-mode curiosities, together with the lively illustrations of such necessary figures, as are referred to practise: approved by the many years experience and carefull industry of *Robert May*, in the time of his attendance on several persons of honor.

18. The exquisite letters of Mr. *Robert Loveday*, the late admired Translater of the volumes of the famed Romance Cleopatra, for the perpetrating of his memory, publisht by his dear brother Mr. A.L.

19. The new world of English words, or a general Dictionary containing the Termes, Dignities, Definitions, and perfect interpretations of the proper significations of hard English words throughout the Arts and Sciences, Liberal or Mechannick, as also all other subjects that are useful or appertain to the Language of our Nation, by *I.T.* & others in *Folio*.

FINIS.

www.ingramcontent.com/pod-product-compliance
Lightning Source LLC
Chambersburg PA
CBHW030446220526
45464CB00006B/2434